T0136414

Cyber-Physical Systems: Decision Making Mechanisms and Applications

RIVER PUBLISHERS SERIES IN CIRCUITS AND SYSTEMS

Series Editors

MASSIMO ALIOTO
National University of Singapore
Singapore

KOFI MAKINWA
Delft University of Technology
The Netherlands

DENNIS SYLVESTER
University of Michigan
USA

Indexing: All books published in this series are submitted to Thomson Reuters Book Citation Index (BkCI), CrossRef and to Google Scholar.

The "River Publishers Series in Circuits & Systems" is a series of comprehensive academic and professional books which focus on theory and applications of Circuit and Systems. This includes analog and digital integrated circuits, memory technologies, system-on-chip and processor design. The series also includes books on electronic design automation and design methodology, as well as computer aided design tools.

Books published in the series include research monographs, edited volumes, handbooks and textbooks. The books provide professionals, researchers, educators, and advanced students in the field with an invaluable insight into the latest research and developments.

Topics covered in the series include, but are by no means restricted to the following:

- Analog Integrated Circuits
- Digital Integrated Circuits
- Data Converters
- Processor Architecures
- System-on-Chip
- Memory Design
- Electronic Design Automation

For a list of other books in this series, visit www.riverpublishers.com

Cyber-Physical Systems: Decision Making Mechanisms and Applications

Kostas Siozios

Aristotle University of Thessaloniki
Greece

Dimitrios Soudris

National Technical University of Athens
Greece

Elias Kosmatopoulos

Democritus University of Thrace
Greece

River Publishers

Published, sold and distributed by:
River Publishers
Alsbjergvej 10
9260 Gistrup
Denmark

River Publishers
Lange Geer 44
2611 PW Delft
The Netherlands

Tel.: +45369953197
www.riverpublishers.com

ISBN: 978-87-93609-09-9 (Hardback)
       978-87-93609-08-2 (Ebook)

©2017 River Publishers

Contents

2   **On Designing Decision-Making Mechanisms
    for Cyber-Physical Systems**   **99**

Kostas Siozios, Dimitrios Soudris and Elias Kosmatopoulos

3   **Design Space Exploration Methodology Based on Decision
    Trees for Cyber-Physical Systems**   **111**

Lazaros Papadopoulos and Dimitrios Soudris

4   **PReDiCt: A Scenario-based Methodology for Realizing
    Decision-Making Mechanisms Targeting
    Cyber-Physical Systems**   **129**

*Nikolaos Zompakis, Kostas Siozios, Lazaros Papadopoulos
and Dimitrios Soudris*

Preface

Recently, the convergence of emerging embedded computing, information technology, and distributed control became a key enabler for the future technologies. Among others, a new generation of systems with integrated computational and physical capabilities that can interact with humans through many new modalities have been introduced. Furthermore, it is expected that computing and communication capabilities will soon be embedded in all types of objects and structures in the physical environment. Applications with enormous societal impact and economic benefit will be created by harnessing these capabilities across both space and time domains. Such systems that bridge the cyber world of computing and communications with the physical world are referred to as cyber-physical systems (CPSs). Specifically, a CPS is a collection of task-oriented or dedicated systems that pool their resources and capabilities together to create a new, more complex system which offers more functionality and performance than simply the sum of the constituent sub-systems. Among others, these new design paradigms have the ability to interact with, and expand the capabilities of, the physical world through monitoring, computation, communication, coordination, and decision-making mechanisms. Such an emerging multidisciplinary frontier will enable revolutionary changes in the way humans live, while it is also expected to be a key enabler for future technology developments.

The integration of physical processes and computing is not new. Embedded systems have been in place for a long time and these systems often combine physical processes (e.g., through digital/analog sensors) with computing. However, the core differentiator between a CPS and either a typical control system, or an embedded system, is the communication feature among systems components, which adds (re-)configurability and scalability, allowing instrumenting the physical world with pervasive networks of sensor-rich embedded computation. The goal of CPS architecture is to get maximum value out of a large system by understanding as how each of the smaller (sub-) systems work, interface and are used. This trend is also supported by the continuation of Moores law, which imposes that the cost of a single embedded computer equipped with sensing, processing and communication capabilities drops to- wards zero. Thus, it is economically feasible to densely deploy networks with very large quantities of sensor readings from the physical world, compute quantities and take decisions out of them. Such a very dense network offer a better resolution of the physical world and therefore a better capability of detecting the occurrence of an event; this is of paramount importance for a number of foreseeable applications. Apart from their efficiency to deploy complex systems, CPS exhibits also substantially stricter performance constraints. This problem becomes far more challenging when (i) real-time constraints have to be met, (ii) entities comprising a CPS run over heterogeneous environments, and (iii) these entities interact with each other in a very complex manner. To make matters worse, existing approaches for CPS usually impose components with increased processing power, which is not always the case especially at the embedded domain. Moreover, as the physical world is not entirely predictable, it is not expected the CPS to be operating in a controlled environment; thus, it must be robust enough to unexpected conditions and adaptable to subsystem failures. Finally, it is also expected that the derived products should be highly extensible for new functionalities that enable flexible adaption especially under run-time or real-time constraints.

A critical challenge for designing an efficient CPS relies on the decision-making mechanism. The majority of existing control mechanisms exhibit increased computational and/or storage complexity, which in turn makes their implementation as part of an embedded system a challenging issue. The importance of this problem was also considered by research institutions and industry as a big challenge for upcoming large-scale CPS platforms. Note that the absence of developing lightweight solutions (able to be executed onto embedded platforms) for supporting large-scale systems decision making is not due to neglect, but rather due to its difficulty. Also, as we have already highlighted, this problem becomes far more challenging in case the decision making has to be made under run-time (or real-time) constraints. In such a case, usually a compromise between the desired accuracy and the processing overhead is performed. Existing approaches for supporting the systems decision making rely mainly on *ad-hoc* methods: After all the components, have been designed and manufactured, the control mechanisms aim at making the system to work somehow. However, as the complexity of engineered systems continues to increase, the lack of a systematic theory for systems decision making introduces additional problems.

This book is at the same time a textbook that provides basic concepts, essential knowledge and course exercises as well as a current snapshot of industrial and academic research in the area of decision making for CPS. It underlines the design challenges and shows the evolution of control theory at this field, in its every facet: methodologies, algorithms, tools, architectures, etc.

The contents of this book follow:

Initially, we discuss an overview about emerging systems-related concepts, approaches, and technologies unifying and advancing achievements of science and technology the past decades. Different type of systems are discussed throughout this analysis, such as the CPS, IoT, I2oT, SoS/E, while emphasis is also given at the cross-cutting decision-making.

The second chapter introduces the alternative solutions for designing an efficient decision-making mechanism targeting cyber-physical systems. Aspects related to the computational complexity, the low-power/energy consumption, as well as the physical implementation of system's orchestrator are analyzed. Additionally, software tools that automated the procedure of system's modeling, simulation and emulation are also provided.

The third chapter emphasizes at the exploration of different architectural-level parameters. This task is crucial especially due to the continues increase of cyber-physical system's complexity. In order to address this challenge, a software-supported methodology able to model efficiently the exploration space in a systematic way is discussed.

After having a detail modeling of the system, it is feasible to implement a proper decision-making mechanism. This is discussed in fourth chapter, where a scenario-based methodology for realizing controlling different part of a cyber-physical system is introduced. For demonstration purposes, a smart building use case is employed, where we highlight the superior of introduced mechanism as compared to well-established control techniques.

Since cyber-physical systems usually are consist of various mission–critical components, where the overall system's security and reliability aspects are crucial, it is absolutely necessary to incorporate techniques that guarantee their proper functionality. These techniques have to be implemented as part of the decision-making mechanism. For this purpose, at fifth chapter we discuss mechanisms that are able to handle the challenges posed by zero-faults (or near zero-faults) specifications. Although emphasis is given at the hardware infrastructure of the CPS platforms, the fault-tolerant solutions are also applicable at software level.

The sixth chapter discuss a framework for research and prototyping in robotics. As these systems require the involvement of many people and many hours of design, development and cooperation; significant time and effort overhead is required for evaluating conceptual ideas in design, control, and technology, and for bringing them fast into reality for testing. The framework's functionality is validated and illustrated by two application examples concerning the control systems of a single-legged hopping robot and an instrumented treadmill.

The chapter seven provides a number of examples, where interested readers can implement components of decision-making mechanisms in software-level. The goal of this chapter is to familiarize the readers with the "control theory", in order to get insight in system design. Without affecting the generality of our approach, regarding the purposes of this book, these examples are provided is Matlab software.

Finally, the last chapter gives an overview about the research and development actions in topics related to the cyber-physical systems and/or decision-making for complex architectures. For this purpose, the outcomes funded under Horizon 2020, European Research Council and Marie Skłodowska Curie actions and discussed.

Kostas Siozios

Dimitrios Soudris

Elias Kosmatopoulos

List of Contributors

Alkis Konstantellos, *1) Senior C&I Engineer, Athens, Greece 2) European Commission (retired), Complex Systems and Advanced Computing Unit, Brussels, Belgium*

Dimitrios Soudris, *School of Electrical and Computer Engineering, National Technical University of Athens, Athens, Greece*

Elias Kosmatopoulos, *Department of ECE, Democritus University of Thrace, Greece*

Evangelos Papadopoulos, *Department of Mechanical Engineering, National Technical University of Athens, 15780 Athens, Greece*

Iosif S. Paraskevas, *Department of Mechanical Engineering, National Technical University of Athens, 15780 Athens, Greece*

Konstantinos Machairas, *Department of Mechanical Engineering, National Technical University of Athens, 15780 Athens, Greece*

Kostas Siozios, *Department of Physics, Aristotle University of Thessaloniki, Greece*

Lazaros Papadopoulos, *School of Electrical and Computer Engineering, National Technical University of Athens, Athens, Greece*

Nikolaos Zompakis, *School of ECE, National Technical University of Athens, Greece*

Spyridon Garyfallidis, *Department of Mechanical Engineering, National Technical University of Athens, 15780 Athens, Greece*

List of Figures

List of Tables

List of Abbreviations

3D	Three-Dimensional
5G	5th generation mobile technologies
a/c	aircraft
ACAS-x	Airborne Collision Avoidance System x
ADS-B	Automatic dependent surveillance – broadcast
API	Application Programming Interface
ARTEMIS	European Platform organisation for embedded computing systems
ASIC	Application Specific Integrated Circuit
ATC	Air Traffic Control (system)
BIP	(Behavior, Interactions Priorities) Methods and language
CAN	Controller Area Network
CA	Collision Avoidance
CAS	Complex Adaptive Systems
CDMA	Code Division Multiple Access
CIM	Computer Integrated Manufacturing
CPS	Cyber Physical System
CSL	Control Systems Laboratory
CSMA	Carrier Sense Multiple Access
Cy	Cyber
DAC	Digital to Analog Converters
DEVS	Discrete Event System (Formalism, methods and specification)
DFA	Deterministic Finite Automaton
DI	Digital input (binary signal)
DM	Decision making
DO	Digital output (binary signal)
DoD/DODAF	Department of Defence Architectural Framework (USA) SoS relevant

DoF	Degree of Freedom
DP or DeP	Decision process
DSP	Digital signal processing
DSS	Decision Support System
EC	European Commission (EU)
ECSEL	Electronic Components and Systems for European Leadership (Partnership)
ECU	Electronic Control Unit (in automatic applications)
EM	Electro-Migration
EoM	Equations of Motion
ERC	European Research Council (EU)
ERP/MRP	Enterprise Resource Planning/Manufacturing Recourse planning
ESA	European Space Agency
EtherCAT	Ethernet for Control Automation Technology
EU	European Union
FAA	Federal Aviation Authority (USA)
FAN	Field Area Network
FCR	Fault Containment Regions
FCS	Frame Check Sequence
FDMA	Frequency Division Multiple Access
FEM	Finite Element Method
FET	Future and Emerging Technologies (part of EU R&D programme)
FIFO	First-In-First-Out
FleXRay	(now ISO standards, ISO 17458) An automotive network communications protocol
FPGA	Field Programmable Gate Array
FR	Functional requirements
FTP	File Transfer Protocol
GAN	Global Area Network
GMA	Gesellschaft Mess- und Automatisierungstechnik (Measurements and Automation Society, Germany)
GPS	Geographic position system
GPU	Graphic Processing Unit
H/D/T	Hydrogen isotopes (Protium/Deuterium/Tritium) used for Fusion energy
h/w	hardware

H2020	Horizon 2020 (European Commission R&D framework 2013–2020)
HC	Hunt–Crossley
HDL	Hardware Decision Language
HiL	Hardware in the Loop
HitL	Human in the Loop
HMI	Human machine interface
HPC	High Performance Computing
HTS	High Throughput Screening (technology)
HTTP	Hypertext Transfer Protocol
HVAC	Heating, ventilating, and air conditioning
HW	Hardware
I/O	Input/Output
I2C	Industrial Internet Consortium
I2oT	Industrial Internet of Things
ICAO	International Civil Aviation Organization (UN agency)
IEEE	Institute of Electrical and Electronics Engineers
IETF	Internet Engineering Task-Force
IFAC	International Federation of Automatic Control
IIOT	Industrial Internet of Things
IMU	Inertial Measurement Unit
INCOSE	International Council on Systems Engineering
IoT	Internet of Things
IP	Internet Protocol
IP	Intellectual Property
IPG	Interpacket gap
IRL	Integration readiness Level
ISA	International Society of Automation (earlier Instrument Society of America)
ITU	International Telecommunication Union
LAN	Local Area Network
LQG	Linear–Quadratic–Gaussian
LSS	Large Scale Systems
LTI	Linear Time-Interval
M2M	Machine to Machine
MAC	Media Access Control
MCS	Mixed Criticality systems
MCU	Microcontroller Unit

MES	Manufacturing Execution System (typical integration Middleware layer)
MiL	Model in the Loop
MoD/MODAF	Ministry of Defence Architectural Framework (UK) SoS relevant
MOSFET	Metal–Oxide–Semiconductor Field-Effect Transistor
MPC	Model Predictive Control
MRL	Manufacturing readiness level
MTTF	Mean-Time-To-Failure
NASA	National Aeronautics and Space Administration
NBTI	Negative Bias Temperature Instability
NCS	Networked Control System
N-F R	Non-Functional Requirements
NIST	National Institute of Standards and Technology (USA)
NM	Nautical Miles
NSF	National Science Foundation (USA)
NTUA	National Technical University of Athens
OMG	Object Management Group
OSI	Open System Interconnection
Phy	Physical
PID	Proportional-Integral-Derivative
PLM	Product Life Cycle management (systems)
PPD	Predicted Percentage of Dissatisfied
PRE	Preamble
PWM	Pulse Width Modulation
QEI	Quadrature Encoder Interface
QoS	Quality of Service
R&D	Research and Development
ROS	Robotics Operating System
RT	Real time
RTL	Register Transfer Level
RTS	Run-Time Situation
S&T	Science and Technology
s/w	software
SAE	Society of Automotive Engineering
SAHR	Single Actuated Hopping Robot
SBC	Single Board Computer

SCADA	Supervisory control and data acquisition (high level architecture/layer/system)
SDL	Specification and Description Language
SDMA	Space Division Multiple Access
SE	System engineering (part of which is SoS E)
SEBoK	System Engineering Book of Knowledge (SE Guide)
SEI	Software Engineering Institute at Carnegie Mellon University
SESAR	Single European Sky ATM Research (Joint Undertaking)
SFD	Start Frame Delimiter
SiL	Software in the Loop
SLIP	Spring Loaded Inverted Pendulum
SM	Stress Migration
SMC	IEEE Systems Man and Cybernetics Society
SME	Small, Medium sized Enterprises
SMTP	Simple Mail Transfer Protocol
SoC	System-on-Chip
SoF	System of the Future (concept)
SoS	System(s) of Systems
SoSE	SoS Engineering
SPI	Serial Peripheral Interface
SRL	System (also Software) Readiness Level
SS	State-Space
SW	Software
TC	Thermal Cycling
TCP	Transmission Control Protocol
TDDB	Time-Dependent Dielectric Breakdown
TDMA	Time Division Multiple Access
TF	Transfer Function
TLM	Transaction-Level Model
TRL	Technology Readiness level
TTA	Time Triggered Architecture
UDP	User Datagram Protocol
ULSS	Ultra Large scale systems
UML	Unified Modeling Language
UPDM	Unified Profile for DoDAF/MODAF (by OMG)

VDI/VDE	Vereins Deutscher Ingenieure/Verbands der Elektrotechnik, Elektronik und Informationstechnik (Engineering Societies – Germany)
VF	Velocity Factor
WAN	Wide Area Network
WEF	World Economic Forum
WSN	Wireless Sensors Network
ZPK	Zero-Pole gain

1

An Overview of Emerging Systems-Related Concepts, Approaches and Technologies Unifying and Advancing S&T Achievements of the Past Decades (e.g. CPS, IoT, I2oT, SoS/E, 5G and Cross-Cutting Decision Making)

Alkis Konstantellos[1,2]

[1] Senior C&I Engineer, Athens, Greece
[2] European Commission (retired), Complex Systems and Advanced Computing Unit, Brussels, Belgium

> *"There is nothing more difficult for a truly creative painter than to paint a rose, because before he can do so, he has first to forget all the roses that were ever painted."*
>
> *Henry Matisse*
> *Painter & Artist (1869-1954)*

> *"The 21st century will reveal even more wonderful insights than the 20th, ... but for this to happen, we shall need powerful new ideas, which will take us to directions significantly different from those currently being pursued"*
>
> *Roger Penrose, Mathematical Physicist (1931-)*
> *"The Road to Reality", 2004, page 1045*

Abstract

In this introductory level overview, the emerging areas of Cyber-Physical Systems (CPS), Systems of Systems and their Engineering (SoS/E), Internet/Industrial Internet of Things (IoT/I2oT), the forthcoming 5th Generation

---

Disclaimer: The opinions expressed in this chapter, the selected project examples and their grouping, are those of the author and do not necessarily reflect the views of the European Commission on the topics discussed.

1

Mobile systems (5G) and cross-cutting Decision Making are discussed with emphasis on their underpinning domains, including control and computing engineering. Several examples illustrate the concepts related to large and small scale systems e.g. mid-air collision avoidance, cryogenics and other instrumentation applications. Beyond historical references, an intuitive profiling scheme addressing {*cyber, human, social* and *physical*} content is proposed. While the origins and the nature of the above-mentioned topics are different, there are complementarities and synergistic combinations. More than just their fashionable names, these areas have opened new opportunities for developing the next generation of high performance, safe and secure systems. However, and despite the progress made to date (2017), *"the expected level has not yet been reached"* (recent NSF comments [1]). We need more than small incremental improvements and due integration of today's design methods, that is, radically new ideas to better understand, model and manage the cyber and physical worlds and their symbiosis – frequently with humans in the middle. Therefore, the investigation of promising mathematical methods, which might come for example, from novel sets and geometric theories, algebras, effective parametrisations, uncertainty quantifications, or advanced signal and logic representations, may be of help. We briefly discuss some of these ideas. This chapter is based on extended current literature and debates across a broad systems spectrum. Furthermore, it includes additional observations, industrial perspectives and personal views by the author, as food for thought.

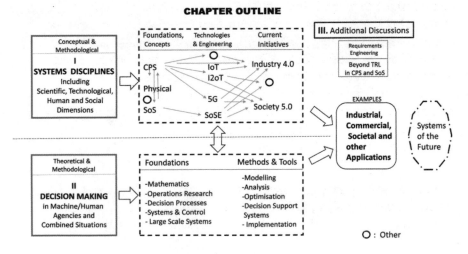

Figure 1.1   Chapter's overview.

Part I – Systems

1.1 Introduction

Ten years after the term and notion of Cyber-Physical systems (CPS) was coined in the US and the almost parallel developments towards the Internet of Things (IoT), an enormous momentum of scientific and industrial activities has been created. Furthermore, from a different starting point, research on special classes of large scale systems, such as Systems of Systems (SoS), Complex Adaptive Systems (CAS) and Ultra Large Scale Systems (ULSS), was necessary to support the sustainable operation and management of such systems. Finally, several interlinked developments have been initiated to enable the realisation of the ambitious 5th generation mobile networks for 2020. The 100+ books and the high number of papers published, is only a small example of the great interest in these promising topics. Big policy initiatives have also emerged, such as Industry 4.0 (Germany), Super Smart Society 5.0 (Japan), Vision-2025 (Korea), Digitizing Europe (EU), Artificial Intelligence and Advanced Manufacturing (US) and China-2025 (China) [2–10]. These major efforts appear as separate and independent paths, with an initial amount of hype and frequent debates among them, but they do exhibit useful complementarities to be exploited, while in practice, each one has an important and distinct role to play in designing, engineering and applying advanced systems in the future.

In addition to Systems and Decision Making (DM), two other topics are discussed: a) end-to-end Requirement Engineering (functional and non-functional) and b) Technology maturity levels beyond TRL, (especially concerning large CPS and SoS) an area where enhancements are needed.[1]

---

[1]Notes: 1) Abbreviations CPS, SoS and IoT are used in singular, or plural. They are called new topics in this chapter; 2) terms like ICT, system, complexity, control, engineering, design, reflect their popular meaning, while more rigorous definitions are presented in the subsequent sections; 3) a few mathematical notations are used, because of their power to express definitions and statements in compact form; 4) *"Physical parts and Physical systems"* are considered e.g. mechanical, thermal, chemical, electrical and other human-made systems, or natural resources; 5) Computer/electronic hardware is a special case discussed under CPS, therefore, some hardware by convention is classified as physical, other as cyber, depending on the context and the relative scale of the systems; 6) Biological and non-human organisms could be placed under physical and 7) important and related topics and technologies such as Cloud/Fog, Big Data and High Performance Computing are only briefly mentioned.

1.1.1 Key Survey, Review, Reference Publications and Textbooks

Since about 2010, there have been several survey papers, concerning the foundations of the new topics, their specific aspects (e.g. security [29, 131, 132]) and a broad set of applications. Furthermore, excellent textbooks are also available today providing comprehensive, fresh and rigorous bases for the "new topics", in system modelling, analysis, advanced design theories and methods. Of high educational quality, also giving valuable engineering insights, are for example the books of Edward A. Lee & Sanjit Seshia, 2nd ed. 2015 [11] and Rajeev Alur 2015 [12] on the foundations and the transition from embedded systems to CPS; Peter Marwedel 2011 [13] covering embedded hardware & software systems and architectures towards CPS, Mykel Kochenderfer 2015 [14] presenting advanced computational methods and applications of decision making under uncertainty, Mo Jamshidi 2009 [17] on SoS and SoSE origins, applications and challenges, Hermann Kopetz 2011 [15] on real-time safety critical architectures and networks, Edward Lee & Pravin Varaiya [18] on signal and systems and A. Fradkov on Cybernetical Physics [19]. Other useful books are included in the references e.g. [16, 77, 78, 130, 154, 193, 207, 293]. There are also several well-prepared and easy to read overview, discussion and state of the art reports and papers, for example the NIST Framework [20], Cyphers, CPSoS and Platform4CPS projects providing CPS overviews and state of the art from an embedded computing and tools point of view [36, 197, 291], and similarly [39] report on SoS/E, [33] on smart agents in production, the IFAC-2017 strategic impact of control report [23], László Monostori paper on production CPS [144], E.Lee on the future of CPS [102], Kyoung-Dae Kim and P. R. Kumar comprehensive CPS Overview [21], Babiceanu & Seker on Big Data in CPS [22], Vermesan & Friess on IoT, Heemels on CPS and Control [32] and useful surveys and discussions [24–28, 33, 37, 39, 41, 43–45, 213, 246]. Constructive criticism has also been published, mainly concerning some overlappings between certain new topics (IoT ↔ CPS ↔ SoS) and the lack of clarity in their definitions [142, 145–153, 290]. We discuss these debates in the relevant parts of this chapter.

Note: A comprehesive list with comments regarding recently funded projects by the European Commission, in the topics discussed in this chapter, are included in Chapter 8 of this book.

1.1.2 A Motivating Example: Air Traffic Management and Collision Avoidance Systems

We start with the air-transportation paradigm which demonstrates well the degree of systems complexity encountered today and in view of forthcoming developments. Air Traffic Control (ATC) and especially Collision Avoidance (CA) problems attracted the attention not only of the aerospace, but also the systems and control communities since the 90's. It was studied as a hybrid dynamic system and control problem, as in the first rigorous modelling and analysis papers [46, 48], and also in [47, 49], (further discussion in Section 1.8.4.3). Since then, the physical problem requirements and formulation became much more demanding with many new uncertainties, difficult to tackle without some approximations. Today, about 15 years later, few old and some new dedicated R&D teams continue their intensive efforts on these challenging problems, now supported also by government (e.g. FAA, SESAR, EUROCONTROL, ICAO) and industrial standardisation authorities. The proposed new approaches are still using e.g. dynamic and mixed programming, Petri nets and versions of Markov Decision processes, but with more practical optimisation and advanced verification approaches with human safety and usability as the prime goals, see e.g. M. Kochenderfer [45–47], J.-B. Jeannin and A. Platzer [54, 55], C. Tomlin [56] and other work [58–61] on these challenges. Beyond modelling and implementation, some legal issues [50] and harmonisation needs between geographic regions have been discussed [57].

But before we endeavour to label this example, or parts of it, as CPS, SoS, or DM, let us describe the overall application:

The forecasted increase of travelling people in the years to come, leads to concrete industrial, commercial and socio-economic targets to meet, under stringent requirements and constraints, such as higher air traffic density, safety, speed and low cost. In Figure 1.2, the main systems and services involved are shown schematically: Air traffic control/management, on board equipment, ground radar stations, satellite systems, ADS-B localisation transponder systems, radio communications, meteorological data transmitters, airport infrastructures, pilots, passengers and other human operators. In the middle three aircrafts (a/c) are pictured in airspace volume V over period $T\Delta$. (The forthcoming ACAS-X system, is further discussed in the Decision-Making Section 1.8.4.3. ACAS stands for Airborne Collision Avoidance, X denotes special classes (e.g. U concerning UAV intrusion).

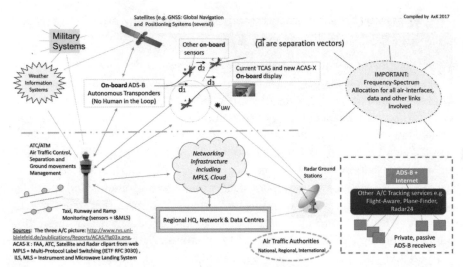

Figure 1.2   Systems involved in tracking, positioning and disseminating situational awareness, Geo-location, weather information for collision avoidance and additional Private Services.

Increased air traffic means: a) that new rules and advanced air traffic management methods are necessary for ensuring safe flights in high-density airspace (i.e. smaller allowable air separation); b) but smaller distances means that planes might fly very close to each other (limit was 100N-Miles in the past, 30NM today and probably <15NM in the future), thus increasing the likelihood of mid-air accidents; c) then, to avoid such collision risks, we need suitable technical systems and procedures, better than the existing ones, to be conceived, designed, developed and deployed, be compatible with relevant infrastructures and radio frequency spectrum allocations and finally d) need to coordinate all systems involved to provide smooth operations under additional encounters, not experienced in the past, such as the presence of hobby and emerging professional Uninhabited Aerial Vehicles-based services (in the same airspace) [62, 63]. Moreover, other useful services with social benefits have also been derived using some of the new tracking technologies, as "read only", beyond and independently from air traffic control systems. Examples include the popular Flight-Aware, Plane-Finder and Radar24 systems and services [68], freely available on the web.

Comment: The multiple system requirements of this example, cannot be addressed and implemented simply as one gigantic project, but require intensive long-term commitments and actions by several systems design and manufacturing industries, aerospace industry, standardisation and governmental organisations and assuming successful R&D activities before commercialisation. Additional discussion on ATC and ACAS is in Section 1.8.1.3 of this chapter.

1.1.3 Success and Failure of Contemporary Systems

Similar challenging situations can be found in many other sectors, such as energy, communications, land and sea transportation, construction industry, health-care, as well as in banking and digital entertainment. In general, in our contemporary life, many technological advances have been established following (industrial and social) behavioural patterns and beliefs regarding innovation. On the other side, malfunctions and serious/faulty operations are vigorously criticised. While the balance is on the positive side, and the number of accidents has been reduced (e.g. in Europe 31% reduction of significant rail accidents between 2006 and 2015), see UIC report and graph, in Figure 1.3 and [75]) the minority exceptions and the continuous demand for improvements amplify the needs and expectations for having better systems in the future. Table 1.1 gives a few examples, representatives of both sides – success and failure. Sources about the examples: [64–74].

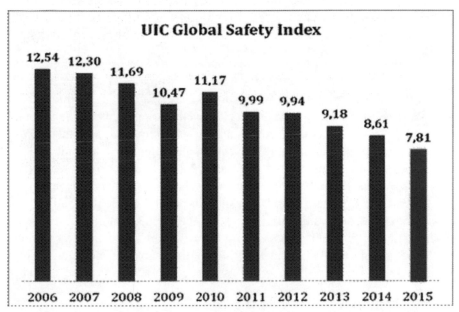

Figure 1.3 Reduction of significant rail accidents, from total 2370 (2006) to 1624 (2015), was monitored in 22 member states in Europe.

Source: UIC, 2016 [75].

Table 1.1   Examples of real life systems design, engineering and management

I. Remarkable systems-related achievements:	II. Some questionable behaviours:
Concepts, Designs, Engineering realisations and human management, using rigorous and effective scientific methods and complex technological innovations, which are proven correct under stringent operational conditions	Likely due to either incomplete requirements capture and interpretation gaps, design difficulties/omissions, lack of standardisation, implementation changes, human in the loop errors, cost reduction, or combinations of these factors
1) 101 Tower, Hsinyi, Taipei and St. Francis Shangri-La Towers in Manila: Tuned- Mass dampers. High buildings anti-vibration, motion control systems	a) Sandiago de Compostela-Spain (2013) and Philadelphia-USA (2015), two similar fatal accidents. Human in the Loop & Safety systems issues
2) Gotthard Tunnel (CH, (2016), Superb sustainable construction engineering	b) Massive recalls of cars and mobile phones [software/ batteries issues]
3) Blade-less Fan [Dyson, UK], Radical non-ICT innovation, Systems modelling and simulations extensively used in the design & manufacturing of the physical system.	c) Remote over-the-air-software updating of ECUs (e.g. Firmware). Induced vulnerabilities regarding car safety and data links security
4) Decision-Support Systems based on rigorous systems and control methods (MDP, POMDP, Games Theories, Dynamic Control and Optimisation, Petri Nets)	d) Long delays in the finalisation of F-35 aircraft (US) also due to technology maturity situations (e.g. TRL mismatch)
5) Free access to (almost global) airplane tracking [e.g. Radar24, Flight-Aware, and Plane-Finder systems and services]	e) Earlier aerospace accidents due also to software and systems
6) Very low-latency systems: for next generation mobile networks such as 5G (≤5mS), digital music synthesis, digital photography and other applications	
7) 3-D Printing and 4-D Printing (including time) with several sectoral applications, including construction industry and agriculture	

Compiled by AxK 1-2017

1.2 System, Model(s), Many Systems and Their Characterisation

1.2.1 What Is a General System?

It appears a trivial, but interesting question. Long time ago, Hall and Fagan [76] gave this compact definition: "*System is a set of objects together with the relationships between the objects and their attributes. This implies that a system has properties, functions and purposes distinct from its constituent objects, relationships and attributes*". In other words, a system is a set of parts and rules describing their interactions. That is, a system has more than one inherent "part", and a "part" is an elementary component for the system. However, a "part", or "component" is not easy to define, because there is **no generally accepted terminology**. In sectoral and scientific conventions, it is useful to promote the use of the same terminology within peer researchers, designers and practitioners, but unfortunately, there is no universally accepted one – "Part" of a system could be called in different domains and industries: component, subsystem, unit, block, object, element, atom, sub-process, group, entity (typical IoT and 5G term), or just "thing". See also Carayannis et al. [105] for innovative considerations of systems from an entrepreneurship point of view.

Formal definitions of a system

To mathematically define a system, we borrow simple definitions, one proposed by M. Mesarovic [77] and one by Jan Willems [78]: "*A general system* *is a relation S, on abstract, non-empty sets X and Y, that is $S \subseteq X \times Y$. If S is a function $S : X \to Y$, then we call S a functional system"*. Alternatively, S may be interpreted *"as a pair of two sets (X, f) where X is the phase, or state space, and f represents the rule(s) to identify the elements of X. If we also add a set T indicating the time instances t, then a system is defined as a tuple $S : (X, Y, f, T)$ (t may be continuous, or discrete). In this case, we talk about a dynamical system"*.

The role and importance of time

While the sets X, Y (called *<inputs* and *outputs>* e.g. in Control and *<observations, actions>* e.g. in Decision Making) are well understood and modelled in most disciplines, representing e.g. physical states and the governing laws (describing e.g. trajectories, velocities, temperatures, numerical and statistical variables as states), time is a more complicated variable, in particular within distributed computer-based systems. Bart Jacobs [79] describes a hybrid system as a co-algebra [80, 81] and time as a monoid action on the state space. Furthermore, Claudio Mazzola [82] defined a dynamical system, in which *"time is only required to satisfy the algebraic features of a monoid, i.e. a semigroup with identity"*. See also [83–88] about time in Automata and CPS where the resilience of the distributed clocks may need special attention.

At a more general level, we observe that in systems and control design, applied e.g. to sequential, parallel and mixed "physical" phenomena, time is essential in their real-time programming and the interfacing with the external world. However, in mainstream computer sciences (except in parts of embedded systems), time is not an obvious attribute, or variable. E. A. Lee [11, 84, 100, 104] Lee and Varaiya [18], Alan Burns [94] and Joseph Sifakis [115, 119] have repeatedly emphasised the importance of time, its inadequate treatment and its implications, in particular, when systems are not monolithic/confined, but distributed, networked, or "open" and need to be subject to different synchronisation requirements. See [15, 19]. The term *"open"* is understood here in the sense of having external communications e.g. via Internet.

The System of Systems and SoS/Engineering communities use the following definitions

System is a functionally, physically, and/or behaviourally related group of regularly interacting, or interdependent elements; that group of elements forming a unified whole [17, 89, 93].

- *Subsystem* is a system that is part of one, or more larger system[2].
- *System of Systems* is a set or arrangement of systems that results when independent and useful systems are integrated into a larger system that delivers unique capabilities [89]. See also some criticism about the SoS definition in [149].

Capability is the ability to achieve a desired effect under specified standards and conditions through combinations of ways and means to perform a set of tasks [93].

All these are high level, macroscopic considerations useful for SoS Engineering and not only.

A System and its attributes

We will consider two general cases of systems, whereby it becomes clear that *time*, if added, indeed complicates the initial static picture: i) a *single system* and ii) *several systems* in an environment.

1.2.1.1 One system

First, we focus on a single system, S_1, as shown in Figure 1.4. It is essential to consider its environment [13, 78]. The environment could be another human-made, computer-based system, or a physical system (mechanical, chemical, thermal, biological, etc.), human operators, a mixture of these, or an empty space. An approach to *identify and model/describe S_1*, is to associate it with a set of attributes, some of which are normally time-invariant, at least for a period, like its tag-number ("identity, or address" to distinguish it from other systems), its operational, or existential goal(s), the number of parts and the number of ports/links. We also distinguish its initial state, its status (e.g. inactive, stand-by, active) in this example. Beyond the basic system attributes, there are systematic and rigorous methods to describe the model of the system, with, or without time considerations, such as *intelligent agents, actors, automata, graphs, or dynamic descriptions* (through e.g. differential, difference equations, differential inclusions, or other relations). Agents are very popular, as distributed software constructs which can interpret e.g. local or networked control logic, be decision makers, mediators, moderators and predictors. Frequently an agent represents a complete local system such as a controller, or broker [296].

---

[2]https://en.wikipedia.org/wiki/System#Subsystem

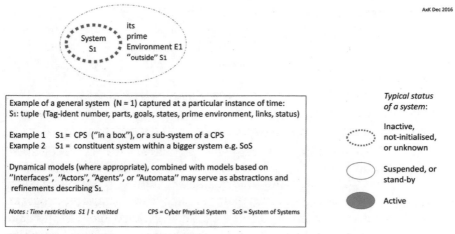

Figure 1.4   A simplified interpretation of a system. For many systems ($N > 1$) refer to Figure 1.5.

Modelling Systems and Time

In Table 1.2, the main methods for first level system modelling are summarised (programming languages omitted). Worth mentioning is that feedback is, in several of these cases, a basic characteristic, but in certain disciplines *"feedback"* is not always associated with a control system.

Table 1.2   Definition of systems and related modeling concepts in various fields

Field / systems: *	Name (Basic)	Typical System definition
Control Theory/* *Control Engineering*	Control Loops (continuous/discrete/ hybrid) *<includes Feedback>*	**Dynamic model** (States, Trajectories) S: (X,Y,S,F, f, T(k) Di, Rj), X: inputs, Y:outputs, S: states, F: Feedback **f: relations between X,Y,S,** T: discrete time, Di: desirable performance, Rj: optimisation rules and constraints
Intelligent Agents/*	Agent, Actor	Tuple (B|Π, S, E), B: Behaviour in perceived Environment Π, S: sensing, E: effectors action
Finite State Machines/*	Petri net, Automaton States & Transitions	Π: (S, Σ, δ, S0, F) , where δ : S x Σ → S the transitions S, S 0, F: States, Σ: triggers
Graph Theory/*	Graph (e.g. directed, Undirected, weighted)	Ɠ (V,E), set of ordered pairs V: Vertices (e.g. states), E: Edges (e.g. control variables) E.g. Control modelling
Decision Making/*	Decision Process *<includes Feedback>*	A: (Oj, Aj, vj, Ij, Si, F), where Oj: Observations, Aj Actions Vj: Evaluations, Ij: Impacts, Si: rankings, F: final selection
Games Theory /*	Game (e.g Non-cooperative)	G(N, Strategy, Pay-offs), N:players
Systems Theory/*	Dynamical System	S: (X,Y,f,T), f: X→Y , T: time (R+ or Z)

Compiled AxK 2-2-17

Note that for each of these methods, their applicability and supporting tools, there exists a vast international literature. The sketch in this figure is a kind of universal structure with optional feedback, which could interpret either a control system, or a decision-making process, or a (closed loop) communication system.

1.2.1.2 Many systems in an environment

At home, single systems are surrounding us, for example a smart hairdryer, a music reproduction set, or a contemporary digital camera. Several of these systems include remarkably advanced high tech subsystems which range from embedded multi-core processors, real time DSP, multi-variable control and estimation algorithms, networking, plus several electromechanical innovations. But the more advanced systems are seldom isolated and stand-alone. They are indeed highly networked. For example, a high-end compact camera mentioned above has today GPS and Wi-Fi links, can control smart lighting, or be remotely controlled and possesses Human-Machine Interfaces (HMI) for operation and software configuration.

Extending the system notion to many systems, in the same, or connected environments, we distinguish several possible situations. Conceptually, but also supported by practical applications, it is instructive to make an abstract categorisation like Figure 1.5. This is by no means exhaustive and may not cover all possible situations and combinations and levels of interactions, but depicts cases of interesting aggregated behaviours. The categorisation is static at time t, or discrete time t_k, thus simplifying real situations where **changes and transitions** may happen in each of the member-elements and between them (models might then include e.g. timed Petri nets variants, timed automata and deep dynamics). Starting from the outer left side in Figure 1.5:

- The broadest class is a Global Systems Environment **S**, called the "Universum" by Jan Willems in his Behavioural Systems theory [78], includes member-elements in different statuses, without further known attributes. It is a random crowd-type set with a vague identification.
- Class **E**, embraces systems in at least active, or known status (not yet called subsystems, because we do not know their level of interaction and functional relationship with others) but which we know that they coexist in E (during a period T). Space may be geographic, geometric, situational, or functional (e.g. all taxis, or ambulances in a city X, every morning).

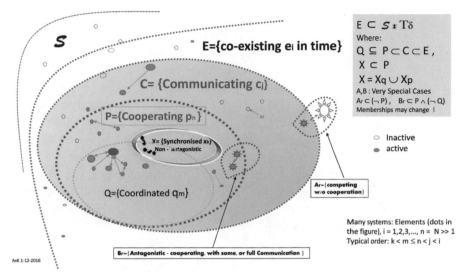

Figure 1.5 Typical functional categories of many objects/entities/systems, initially belonging to a superset S over time period $T\delta$.

- Class **C** refers to systems which, within a subset C of E communicate with each other, or have the capability to interact with each other by data exchange.
- Class **P** refers to elements which beyond communication, cooperate with others, for example by aligning certain actions/decisions to accepted framework, but no central control is assumed.
- Class **Q** is the cross-section of P and C, refers to coordinated elements, which do stronger cooperation under tighter streamlining e.g. goals/trajectories/exceptions logic/timing. May be self-organised, or under separate coordinating authorities(s). Could have a hierarchical structure.
- Class **X** refers to fully synchronised elements either in time, or functionally or both. Note that some elements synchronised in time (e.g. clock synchronisation via the communications capabilities of the elements), may not be functionally synchronised under the same mechanism, or network. Example: in distributed control, where the processors clocks are synchronised periodically through a dedicated network, however the correct coordinated sequencing of the functional software blocks is usually achieved and monitored by a different method [15, 16, 40].

- Special classes **A** and **B**: A refers to situations of competing, non-cooperating elements which either regularly communicate, or spontaneously change role and intervening within, or from outside class C. Typical examples can be found in competition phenomena in genetics [106–111], mobile base station, product versioning and typical mid 90's, PLC vs DCS and vs SCADA control systems offered at commercial level competitive biddings. Similarly, B refers to systems which compete while cooperating within a mutually agreed set of rules, or after a consensus building action. It is frequently the case of bi-polar strategies [112, 114]. An interesting example can be found in hybrid supply chain management as discussed in [106], it is also encountered in biology, when species would have an incentive to coordinate their behaviours, adapting in response to each other's strategies [113, 114].

1.2.2 System Characterisations – Elementary Abstractions

We present two characterisations based on: **attributes** and the **nature** of the systems.

1.2.2.1 Characterisation through fundamental attributes

The single system example of Figure 1.4 was described, by seven attributes (in the frame within the figure). To provide a first, high level abstraction of a general system, a characterisation could be based on only *three basic attributes* as in Figure 1.6. See an early diagram proposed in [241]. While (S_c, B_e, S_t) may not always be strictly orthogonal/independent, they offer a simple universal characterisation tool:

- S_c: Size, B_e: Behaviour, S_t: Structure

$S_c(size, or scale)$: is in most cases independent from behaviour and structure and could intuitively represent a first complexity metric C-I. The size of a system reflects the number and the extend of its subsystems (or parts), the number of sensors and actuators connected (inputs and outputs) and the size of software used. There are however cases, concerning new designs, where a desired behaviour dictates the size of a system, but the (S_c, B_e, S_t) dimensions can still be used.

$B_e(Behaviour\ and\ required\ prime\ function)$: It represents the second complexity metric CII and can be described by mathematical logic models, graphs, agents, actors, games, dynamic systems, or other relations and be parametrized and optimised to meet system objectives and operational goals.

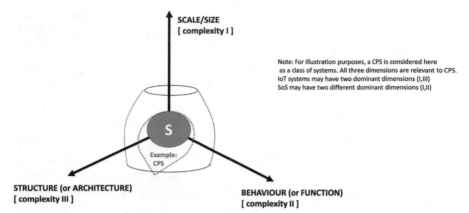

Figure 1.6 Three intuitive dimensions (I, II, III) qualitatively indicating a system's broad characterisation and complexity, based on [241].

A short definition of "behaviour" is $B_e = service \cup failure$, meaning that correct behaviour of a system is when it meets specifications and provides the desirable service, faulty in any violation of specifications, see GENESYS Cross-Domain Reference Architecture [247].

S_t(*Structure and prime Architecture*): is the last dimension in the diagram. It represents the third complexity metric C-III. Structure of a system is partly dictated by the physical design which is assumed fixed. Therefore, the structure of the cyber part (monitoring, controlling, supervising, protecting, advising, explaining, etc) can take the physical part structure as constraint and must comply with best practices and standards to meet operational and other specifications (functional and non-functional).

Remark 1: Despite several known software complexity metrics [123, 124], to our knowledge, none is yet available for integrated CPS and SoS applications. At a more general level, the three axes of Figure 1.6 correspond, intuitively, to three dimensions of *system's complexity* C-I, C-II and C-III, representing some kind of metrics. The R&D communities need to design complexity indicators for future systems.

Remark 2: While structure may imply certain behaviours, architecture does not imply function of a system and size/scale do not imply structure, nor function. However, sometimes ultra large size may imply degrees of complexity and thus certain behaviours. See Sifakis [227] for a comprehensive analysis and discussion concerning the behaviour of systems.

<u>Comment:</u> A system's behaviour can be e.g. linear/non-linear, predictable/ unpredictable, pre-programmed, unknown, deterministic/non-deterministic. Behaviour may also be described by numerical options/outcome-values for example, regarding decision processes, or by constraints. Behaviour will be better understood if we relate it with the prime function of the system. Beyond software-based system considerations, several other disciplines deal with the behaviour of machines and humans and their interaction (e.g. cognitive and social sciences, learning, psychology, management and economics). Such behaviours can still be included in the diagram under e.g. "human in the loop". Emergent behaviour will be briefly mentioned in the System of Systems discussion and is an interesting topic that touches upon several complexity theories.

1.2.2.2 Characterisation according to the nature of a system

Beyond the attempt to classify systems by three simple characteristics (S_c, B_e, S_t) as in Figure 1.6 and in view of the new topics discussed in this chapter, it is appropriate to position the related systems across four additional macroscopic **content spaces (pillars, or attractors)** in Figure 1.7 denoting the nature of the systems (C, H, S, P). This additional characterisation

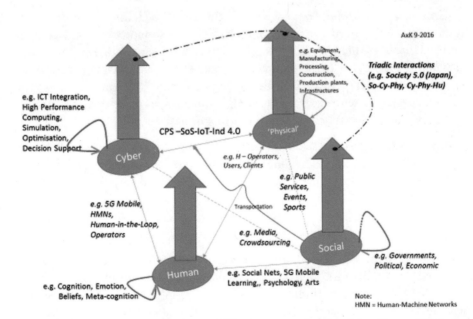

Figure 1.7　Four macroscopic spaces (pillars) representing the Cyber-Human-Physical-Social systems, environments and communities.

is supplementary to the previous one and can be applied to any current, or future human-made system: Cyber, Human, Social, Physical as a linear combination. See [133, 141] and [245] about societal aspects of CPS.

1.2.2.3 An intuitive characterisation-profiling scheme for systems

Defining a) the nature of a system by F: combination of (Cyber, Human, Social, Physical) content and b) the three main system characteristics by a vector S: triple (Scale, Behaviour, Structure) we can compose a general qualitative characterisation index C for systems as: $C : (F, S)$.

Example 1: Considering ACAS-X as a large CPS: $F = (C, H, \emptyset, P)$, $S =$ (Ultra Large, DAL-4, Distributed/Human in the Loop).

Example 2: An ADS-B passive flight tracking service, will be characterised by the following composition: $F = (C, H, \emptyset, \emptyset)$, $S =$ (Large, DAL-1, IP/crowdsourcing). This characterisation C may be useful in macroscopically comparing systems. This approach is customised for a general CPS as a 6-axis radar diagram in Figure 1.15 in the CPS section in this chapter. On the same diagram the ACAS-X example is superimposed as illustration[3].

1.3 Specific Systems Classification in Established R&D-S&T Databases

1.3.1 Condensed Outline-Definitions of the New Topics and Initiatives

- **CPS**: proactive, systems-oriented umbrella concept, gradually becoming an overarching class of systems, which deeply integrate cyber and physical parts and unify a set of underpinning S&T domains, as defined in e.g. NSF, NIST, ECSEL, ARTEMIS.
- **IoT**: emerging networking technology of (mainly massively) deployed objects, sensors, actuators and sub-systems usually based on IP, being extended to an Industrial Internet platform, as defined in ITU, IEEE IS, IETF, H2020 and I2C.
- **SoS**: special class of large scale systems (set, or families of systems), usually complying with concrete organisational, managerial and operational criteria towards common goal(s) (MITRE, DoD, CPSoS).

---

[3]Note: DAL stands for "Development Assurance Level" used in aerospace systems, like SIL in other industries. See for example https://www-users.cs.york.ac.uk/tpk/nucfuture.pdf

- **SoSE**: engineering of System of Systems based on Systems Engineering (SE), from Organisational and Computing/Networking/Control/Interoperability points of view, supporting modelling, analysis, and management (INCOSE, IEEE SMC, OMG).
- **Industry 4.0 (Germany, 2013)**: substantial technology oriented initiative, incorporating CPS and IoT as essential technologies, to support manufacturing and other sectors (ACATECH, VDI/VDE).
- **Society 5.0 (Japan, 2015)**: broad socio-techno-economic initiative, emphasising social wealth, comfort, quality of live, services, human participation and contribution via super-smart systems and ICT [5, 6].
- **Vision-2025 (Korea, 2016)**: long term plan covering R&D and societal benefits, Cyber-security [7].
- **China-2025 (China, 2015)**: Major initiative aiming at the manufacturing industries, addressing CPS [9].
- **5th Generation (5G)**: emerging wireless mobile technology (by 2020), for denser cellular connectivity, higher bandwidth, low latency, supporting growth of IoT, big data, distributed media (ITU, GSMA).
- **HPC**: Comprehensive domain in High Performance parallel processing technologies, including developments of multi/many core processors, essential for data centres, large scale simulations, clouds and decision making (HIPEAC, H2020) [248, 249].

Note: The above-mentioned topics include several underpinning technologies. Some are common to all (e.g. h/w platforms, software engineering, control engineering, embedded systems, simulations, networking), others are more relevant to individual topics (e.g. Enterprise architectures, middleware) and others are still under development [8, 10].

1.3.2 Sciences, Technologies, Industry and Related Communities

We distinguish a thematic system classification from the nowadays popular *document identifications* like Dx/Doi (Digital object Identification and tracking, by the DoI Foundation) and from the unique device address names as the Internet/web URL.

Driven by different objectives, specific systems are *classified* into classes, a hierarchy of sub-classes and sub-sub-classes, for example a) by scientific communities such as IEEE, IFIP, IFAC, ACM, FIPA, OMG, INCOSE, b) by several organisations into systematic databases, updated periodically (UN, UNESCO, OECD, WEF, NACE2, SIC, ISCED-education, WPO & regional patent-offices) and c) R&D funding agencies such as NSF, NIST, DARPA (US), ERC, H2020 (Europe), DFG (Germany) [125–128].

Concerning the new topics, almost any scientific event nowadays includes a session on CPS, IoT, big data, and the policy initiatives e.g. Industry 4.0. However, as far as official indicators and databases is concerned, none of these is yet mentioned. For example, while OECD includes IoT, in the top 6 technologies in their 2016 report, the term is not included in their Frascati classification [129]. Reasons for that: i) non-R&D classifications mainly refer to products and services, ii) the updating of the lists takes several years and iii) some of the new systems are classified under traditional topics, e.g. CPS and IoT under "computing devices", "measurement instruments", "communication equipment".

1.4 Evolutionary Paths towards (CPS, IoT, SoS/E) and (Ind 4, Soc 5.0)[4]

1.4.1 The S&T Paths Jointly Leading to CPS

There have been three main developments (1995–2005) which contributed to the synthetic genesis of CPS: a) a cross disciplinary cooperation between the computing and control communities, b) the catalytic role of funding agencies, primarily NSF/CISE in the USA and the intensive activities in embedded systems, supported by the ICT programme in Europe and c) the industry demand for effective system design and execution approaches. All these led to the smart name Cyber-Physical Systems (CPS). This evolution is schematically shown in Figure 1.8 (from top to bottom–arrows 1, 2 and 4).

The first path (1st arrow) is the **part of computer sciences** dealing e.g. with advanced concurrent software and systems, formal methods, rigorous model checking, verification, model-based engineering, Petri nets, discrete events, hybrid systems, abstractions and advanced code analysis. The second path (2nd arrow) is coming from **control theory and control systems**, which evolved from early time fundamental control theories and engineering methods. This area also addresses high level control via suitable middleware, including cloud and edge architectures with links to SCADA, MES, ERP. The third S&T path contributing to CPS (4th arrow from top in Figure 1.8) is the **hardware field of microelectronics and microprocessors** which led to two main and parallel worlds, the PC and the embedded system. Further advances led to "*Networked embedded and control systems*".

---

[4]Note: We call path the *topics* involved and the activities of the *respective S&T communities*.

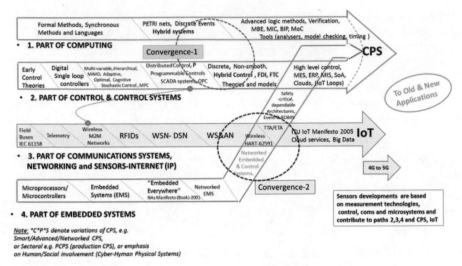

Figure 1.8 Simplified Evolution Path towards CPS (top) and IoT (middle).

Top two arrows: 1 and 2	Two bottom arrows: 3 and 4
MBE = Model-based Engineering	IEC = International Standardisation (el)
MIC = Model- Integrating Computing	M2M = Machine to Machine
BIP = Behaviour, Interactions, Processes	RFIDs = Radio Frequency Identification
MoC = Models of Computation	WSN = Wireless Sensor Networks
MPC = Model Predictive Control	DSN = Distributed Sensor Networks
SCADA = Supervisory Control & Data Acquisition	WS&AN = Wireless sensors & actuators Nets
OPC-UA = OLE for Process Control- Universal Architecture	TTA/ETA = Time/Event Triggered Architecture
FDI = Fault Detection Identification	ITU = International Telecoms Union
FTC = Fault Tolerant Control	IP = Internet Protocol
MES = Manufacturing Execution System (Middleware)	EMS = Embedded Systems (incl. h/w and s/w)
ERP = Enterprise Resource Planning	h/w = hardware
MIS = Management Information Systems	s/w = Software
SoS = Service oriented Architectures	ROBIN = EU project on Event- B methods
CPS = Cyber-Physical Systems	HART = Highway Addressable Remote Transducer using FSK
IoT = Internet of Things	4G to 5G = 4th to 5th generation mobile communications
OLE = Object Linking and Embedding	

Figure 1.9 Abbreviations for Figure 1.8.

An interesting cross-disciplinary cooperation took place in late 90' and continues today: The Computing and Control communities worked jointly on challenging problems posed by the so called "*Hybrid Systems (HS)*". HS is probably one of the main contributors to the concept of CPS. Another highly important S&T area of CPS deal with "*safety critical systems*". Here are two examples of attractive for industry design approaches:

1. **Time-Triggered Ethernet** applies to communication systems with mixed time criticality sharing a single physical network (i.e. synchronous and non-time-triggered modes). It uses an inherently fault-tolerant, self-stabilising clock synchronization algorithm and SMT-based[5] network-level scheduling. Recent applications include the NASA Orion Mission[6]) and the TTA protocol in the Boeing 777 Dreamliner passenger's aircraft. Furthermore, the synchronisation approach has been standardised by SAE as AS6802, 2016[7] [239].

2. **Event-B**, according to http://www.event-b.org/, is a formal method for system-level modelling and analysis. It uses set theory, specific refinements to represent systems at different abstractions and proof based verification techniques to ensure consistency between the respective levels. Its implementation uses the open source RODIN Platform based on Eclipse. These developments have been supported by the E.C. funded projects e.g. RODIN (2004–2007), DEPLOY (2008–2012), ADVANCE (2011–2014) initially led by Jean-Raymond Abrial, at the University of Southampton (U.K.) who proposed the Event-B method several years ago. An industrial tool, called Atelier-B[8] was developed by ClearSy[9] used today in railways safety critical systems and in microelectronics manufacturing[10] [238].

Note: While physical sensors and actuators are integral part of any closed loop control system, (comprising minimum one sensor (real, or virtual) and probably also one actuator), we included them in the electrical, electronics and microelectronics S&T, between embedded electronics, communications and control.

1.4.2 The S&T Paths Jointly Leading to IoT

The IoT roots have been in the first wireless instruments in the early 70s, RFIDs of the 90's, followed by WSN and WS&AS around year 2000. The path to IoT shown on the same Figure 1.8 (3rd arrow from top) started

---

[5]SMT stands for Satisfiability Modulo Theories

[6]https://www.tttech.com/markets/space/projects-references/nasa-orion/

[7]http://standards.sae.org/as6802/

[8]http://www.atelierb.eu

[9]http://www.clearsy.fr

[10]https://en.wikipedia.org/wiki/B-Method

with the effective interaction of several domains: e.g. communication systems technologies, sensors, wired interfacing protocols and distributed sensor networks.

The emergence of wireless sensor networks (WSNs) was a significant development, both conceptually and as a new networked systems architecture. Their debut, as tiny wireless sensors (e.g. Berkeley massive "smart dust" and the popular in the U.K. "smart rocks"), attracted the attention of industry, because of their *low cost, small size, and their capabilities of ad-hoc operation and self-x/self-configurations.*

The path to IoT crosses the path to CPS at the *point* where "*networked embedded and control systems*" meet and partly overlap with "*Wireless Sensors and actuators*". The expanded IoT, as technology and infrastructure intending to include also actuators, imply "closed loops" over networks, see [131, 244].

1.4.3 The S&T Paths Jointly Leading to SoS/SoSE and Independently to Industry 4.0

The birth of System of Systems (SoS) concept and the related System of Systems Engineering field (SoSE), have their roots: a) in the challenges encountered when dealing with large enterprise architectures, organisation, control, quality and computational systems engineering (SE) and b) also in the system problems and software challenges encountered in large defence procurement project.

SoS is a special class of large scale systems and is close to similar terms and concepts, such as Complex Adaptive Systems (CAS) e.g. at Sandia Labs, Ultra Large Scale Systems (ULSS) at SEI at CMU and Global Systems (GS) used e.g. in the EC, FET program. The SoSE community is organised primarily around INCOSE, the IEEE Systems Council and the Systems, Man & Cybernetics society. There have been interesting debates in defining the relative positioning of SoS vs. CPS and other systems and about the IoT-CPS interactions.

Adopting SoSE in practice, frequently presents some extra design and long term burden, but has many benefits too. It is up to the systems owners and designers to evaluate the techno-economic merits of such an extra layer in their plans vs. alternative approaches such as based only on large scale software engineering, or service oriented architectures (SoA). We consider, that SoSE is closer to practical system implementation and standardisation of specific methods and tools (e.g. DODAF, MODAF, UPDM), than the alternative approaches, but all of them have an individual strength in support of deep understanding, modelling and resolving large systems design and organisation chellenges.

Figure 1.10 shows the combination of the key topics and domains contributing towards SoSE and some of them also playing a role in the portfolio of Industry 4.0 undertakings. On the other side, not all of Industry 4.0 elements are necessarily relevant to SoSE (e.g. robotised production).

Industry 4.0 has several roots in the classic Computer Integrated Manufacturing (CIM) initiatives, activities and related R&D efforts of the 70's to 90's, and is based on subsequent CIM advances. These include for example, enterprise systems integration, shop floor and service robotics, Computerised Numerical machines (CNC), broad sensors and actuators technologies, human factors engineering, ergonomics, dependability, automation, control engineering, simulation technologies and currently embraces the emerging IoT and CPS topics as well as, decision making support systems and management and productivity tools. Industry 4.0 also extends earlier initiatives, such as Factory of the Future (FoF), International Manufacturing Systems(IMS) and Advanced Manufacturing Programs. While it was conceived, launched and promoted initially in Germany [44, 134], it has today a broader S&T appeal and international recognition [142–144].

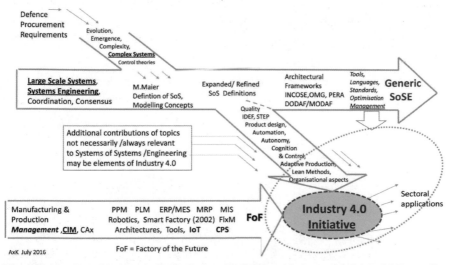

Figure 1.10 Simplified Evolution Path towards SoSE and Industry 4.0 (Main S&T contributing topics are underlined).

1.5 CPS in More Detail: Definitions, Challenges, Debates and Synergies

1.5.1 Introductory Examples

Example 1: Car Engine – Electronic Control Units (ECU) as *local component-level* CPS

A contemporary car has several dedicated electronic control units (ECUs), each of them associated with a monitoring, or control task, such as breaking (ABS), engine management (EM), motorised parts in doors and chassis, steering, stabilising systems (e.g. ESP), condition supervising and the overall coordination of functions. A typical mid-range passenger car can have up to some 60 ECUs, networked through a suitable industrial network such as CAN and fault tolerant architectures like TTA and FLEXRAY. *Note: Based on multi/many core processors and advanced partition or mixed criticality methods, the number of ECUs may be reduced in the future.* The "physical car engine", and each "ECU box with the control system in it", is a local Cyber Physical System with the human (driver) in the loop. Before the introduction of the term CPS, the car electronics ECUs were called a "networked embedded and control system".

Example 2: Elementary Cryogenic instrumentation as *local sensor-level* CPS.

First, let us consider an electrical diode, in isolation. This is a dual use electrical component leading to two, totally different usages and models, an electrical and a thermal (other than the thermal effect of an electric diode). Both are useful in totally different applications.

Same physical component, different behaviours, functions and models

1. A diode's best-known use and prime function is rectification (not measurement). It is a (passive) nonlinear component, but not a system. If we build a bridge with four similar diodes and supply current from a source to its two ends, we get a full cycle rectifying circuit commonly used in AC/DC converters. The bridge is a small sub-system, but not a CPS. A diode could also constitute a subsystem acting on high frequency signals e.g. as part of a demodulation in classic radio receiver systems. The electrical behaviour of a diode is described by the physical law $I(v) = I\alpha \times [e^{(\frac{v}{\eta})-1}]$. A simple approximation is a model by a piecewise linear function: $I(v) = \frac{1}{(2 \times k)}[\nu - \lambda + |\nu - \lambda|], k > 0$ for silicon diodes, with approx. cut-off $\lambda = 0.5$V and is valid for $\nu \geq 0$. A diode (and a bridge) are fixed electrical and electronic components of physical

systems, or sub-systems, but not strictly speaking CPS, because there is no software and no computation involved. The function of a diode can of course be easily implemented in software. Now a second totally different function of a diode.

2. A silicon diode can also be used as a low temperature sensor [135, 136] for cryogenic applications as illustrated in Figure 1.11. This interesting behaviour of the physical diode follows a different (than the rectifying one) piecewise characteristic with excellent sensitivity at very low temperatures. If this diode, acting here as sensor, is connected to a current source and the voltage across it measured by a computer-based system, usually through a suitable electronic converter, could be the basis for controlling a cooling loop. Then we can talk about an *elementary CPS*. It is now a matter of convention where the actual Cy-Phy boundary is.

More general than the specific cryogenic applications, process control plants have many physical measuring and actuation components, say 100's–1000's of different sensors and many reconfigurable physical processes, monitored and controlled by computer systems. In such cases, we can talk about a large-scale CPS. Here the 1st generation CPS is to great extent identical with a contemporary process control system for a complex process. This includes sensors, controls, SCADA and higher levels of MES to ERP with cloud-based links. Such plants (but not many) around the world, "*offer realistic Industry 4.0 paradigms ahead of current developments.*"[11]

Note: 1st generation CPS reflects here the currently available systems (2016).

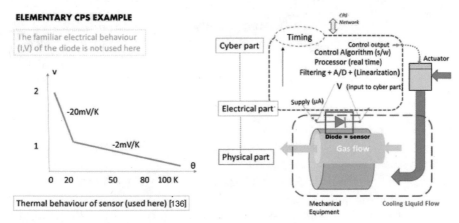

Figure 1.11 An electric diode (Si) is a suitable low temperature sensor for cryogenic applications. The measurement-control arrangement is an elementary Cyber-Physical System.

---

[11] Hannover Fair, Manufacturers comments 2016.

1.5.2 The Success of the Term CPS and Its "Externalities"

The term CPS was coined first in the US around 2006, by Dr Helen Gill of the CISE directorate of NSF [11]. CPS evolved rapidly as a national priority in the US supported by NSF, DARPA, NIST and other agencies. Similar activities were also initiated by the European Commission and individual countries and various organisations giving slightly modified definitions and content to the original CPS. All these efforts aimed at radically upgrading the embedded systems R&D orientations, to embrace future elements, such as the rapidly evolving internet connectivity, to encourage effective rigorous modelling methods for real-time systems and provide the bases for the next generation system architectures.

The term CPS, despite its inherent generality, and some initial hype, has been successful more than alternatives discussed at that time, such as *"Computational Intelligence"* [156], *"Ambient Intelligence"* [155], *"New or 2nd order Cybernetics"* [154], *"Ubiquitous and Pervasive Computing"* and *"Physical Computing"* [157].

One unique attraction point and strength of the term CPS, is the emphasis on the *system*, not implied by the other proposed terms. The high impact and swift adoption of the term CPS around the world, is also due to its role as a unifying attractor and catalyst to diverse domains, thus offering a simple but unique neutral space for many to creatively contribute towards new challenges (under CPS) in a joint manner.

1.5.3 CPS – Definitions

Lee and Seshia [11] define a model of a physical system as *"a description of certain aspects of the system that is intended to yield insight into properties of the system. A model is itself a system"* and that *"Models of CPS (i.e. physical sub-systems together with computing and networking) normally include all three parts"*.

A model of a system (existing or to be built) is useful, (if good) mainly because it facilitates (for example in simulations) making predictions and analysis of the impacts of variations of e.g. structure, parameters and signals on the system's behaviour and the interactions with the system's environment. Edward Lee explains in detail the differences of using a model in sciences and in engineering [11].

A collection of CPS definitions follows:

- Initial NSF Definition [2, 24]: "Cyber-Physical Systems are integrations of computing and physical systems".

- E. Lee and S. Seshia definition [11]: "A Cyber-Physical systems is an integration of computation with physical processes whose behaviour is defined by both the cyber and physical parts of the system" and "CPS is about the intersection, not the union, of the Cyber and Physical parts".
- VDI/VDE/GMA/Industry 4.0 (Germany) (Free translation): "CPS is a system which couples real (Physical) objects and processes with information processing (virtual) objects and processes over information networks which are open, partly global and anytime linked with each other", followed by a remark: "Optionally, a CPS uses local, or remotely available services, possesses Human-Machine Interfaces and has the capability of timely dynamic matching of the system" [137, 138].
- "*How other domains define CPS*" (e.g. IEEE IoT report 2015, see Section 1.6.2 below): Cyber-physical system (CPS): A system of collaborating computational elements controlling physical entities.
- NIST definition: "Cyber-Physical Systems or smart systems are co-engineered interacting networks of physical and computational components. These systems will provide the foundation of our critical infrastructure, from the basis of emerging and future smart services, and improve our quality of life in many areas"[12].
- And probably the best entertaining definition by UPenn "CPS*: Systems that integrate control, computation, and communication (and) can do cool things, and useful things*".

Since a definition cannot be smart, general, future oriented and at the same time specific and crisp, CPS could sufficiently be defined by any of the above simple sentences, if accompanied by a small set of qualifications in the spirit of tight integration of Cy and Phy parts, which should imply the added value and novelty in the design and/or its intended function. For example: use the NSF, VDI, or NIST definition together with a set of differentiating qualifications e.g. "Systems Designed for high confidence use, under cooperative, or ultimately concurrent functional and non-functional requirements considerations, to be correct by construction, integration and evolution and providing high resilience and immunity to external attacks".

CPS Combinations, extensions, derivatives and other aspects

During the last decade, a plethora of derivatives and extensions of the term CPS have emerged, as *C*P*S*, where one of the * is a qualifier, for example

---

[12]https://www.nist.gov/el/cyber-physical-systems

smart, human, social, industrial, networked, production, medical, local, component. Some of these are meaningful (e.g. Cyber-Human-Physical Systems, component CPS, CCPS for Cloud CPS), some others rather superfluous (e.g. intelligent CPS). Further combinations include CPS and SoS, as the E.C. funded project DYMASOS [197]. See also a study [133, 280] on ethical and other issues of CPS.

R&D in CPS is and should not be confined only around known topics, like the ones mentioned above, which rightly need to continue their efforts, but ideally it must include new ideas and methods not available today and unify them under a common extended interdisciplinary field.

Main Aspects, Methods and Tools to consider

Figure 1.12 summarises the three main things CPS designers normally consider: *aspects*, *methods* and *tools*. The entries in each of the three columns reflect the multitude of considerations and the variety of tools and methods a designer should evaluate, select and integrate for a specific application aspect. Certainly, the available commercial methods and solutions have their own target audiences and provide help in the tuning, or configuration, but from the user side, there are also strategic choices to meet at corporate level, or on a case by case basis. Note: the discussion of each topic in Figure 1.12 is beyond the scope of this introductory overview, see [36].

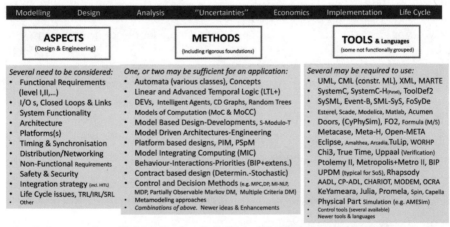

Figure 1.12   Different layers of abstraction for which **system owners** and the designers need to consider several **aspects**, adopt some **methods** and use couple of **tools**.

1.5.4 What Is Not a CPS?

With the initial broad coverage and the current proliferation of the term CPS, penetrating all industries and applications, most of the software-based systems are CPS. This could be an extrapolation of the older statement that "98% of all processors produced go to embedded applications and most embedded systems do link to the physical worlds". There are certain exceptions though. As Janos Stzipanovits has commented [139], that "systems which do only computation are not CPS" and we could add, for example, that these may be isolated (and not networked) desk top PCs, servers, databases etc. Nevertheless, there are also difficulties where to classify some systems such as simulators, decision support systems, and special cases concerning the physical part, like smart phones (which can remotely control home appliances and get real time wheather reports or control toy-drones)".

1.5.5 The Same System Can Be Viewed under Different Perspectives and Modelled through Different Methods

The air traffic example discussed in the introduction and shown in Figure 1.2, including the mid-air collision avoidance system ACAS-X, can be envisioned and classified, at least theoretically, as CPS, DM, or SoS:

- as a large CPS (for design, verification and manufacturing), incorporating distributed model-based systems, where the software would be combining historical big data and real time updating of the a/c position based on on-board sensors and output information on cockpit displays. Dynamic Coordination or consensus aspects may also be important.
- as a real-time decision support system (DSS) based on a lookup table with an advisory type output to the pilots and as safety warning and interactive advisory message. See additional discussion of the ACAS-X system (as DM/DeP) in Section 1.8.1.3.
- as a System of Systems (SoS) where the constituent systems will include e.g. the a/c themselves, the ADS-B transponders constellation, the new ACAS-X on board of a/c, ground radar stations, satellite positioning support with emergent phenomena for example: meteorological data, presence of more than 2 "intruder a/c" near each one "test a/c" and the presence of UAVs. One of the few papers linking SoS and ACAS-X is Daniel de Laurentis, 2015 [140].

Crane Movements: Trajectories under restrictions-based control & Decision Making (within planning)
Simplified state transitions as illustrations of the complex execution in space and time.

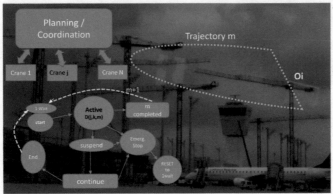

Figure 1.13 Collision avoidance in groups of multiple cranes in large construction sites (airport extension project shown).

Collision avoidance (CA) problems and related R&D challenges to build systems which prevent accidents, are encountered in several other applications such as land transportation, marine traffic and in the construction industry and seaports, (loading terminals). Figure 1.13 presents a simplified collision avoidance logic in multiple fixed cranes configurations in dense construction site, such as in typical airports expansion projects. See also Table 1.4 for other industrial CA applications in the DM Section 1.8.4.3.

1.5.6 The Boundaries between Cyber and Physical Parts of a CPS System

Figure 1.14 schematically illustrates the typical options of the cy-phy boundary within a CPS system. In this context, we should mention slight differences within the various systems communities:

- Within the IoT community, Cyber is called Virtual (e.g. air interface, network protocol, coding decoding); and Physical is called Physical (e.g. modems, sensors/actuators, transceivers, user devices, processing nodes).
- In Control Engineering, Cyber is understood to be the control strategy, the regulatory and optimisation algorithms, programming and tools, systems & applications software; and Physical is the instrumentation and electrical systems "out there" e.g. transmitters (include sensors), transducers (converters), actuators and other final control elements,

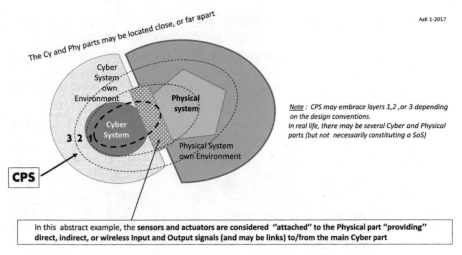

Figure 1.14 Distinguish between Physical and Cyber parts.

and the mechanical, machinery, process/analytical and high/low voltage electrical equipment and units.

- In CPS, cyber could be described, in simple words, as the software-based parts of the system together with the digital electronics & processors/computers and ICT-based sub-systems, which are functionally necessary to execute the required system programmes and logic and implement all applications related to the physical part. Physical on the other side, refers to all human-made parts which play a concrete functional role in the overall application environment called in brief "user system".

- Finally, within CPS theoretical groups, as is the case for example also in control theories, Cyber also includes the theoretical methods and approaches used for modelling both cyber and the physical parts and Physical is the set of models of the physical systems, even if they may be in simulation (Model-in-the-Loop, Hardware-in-the-Loop).

1.5.7 Cascading & Nesting of Multiple Cyber, Physical and Complete CPS

Small scale CPS, called **CPS components** [20], may be aggregated into local CPS and many CPS can further be combined into a large CPS. It is envisaged that there are several applications where an elementary CPS, or more precisely a CPS component (e.g. an ABS controller in an overall car system, or a high Megapixel camera DSP module in a smart phone) is combined with other physical systems which may in turn be equipped with their own CPS and so on, up to the first fully fledged self-contained and identifiable CPS driven product.

This is also the case of several physical systems which are designed with local built-in dedicated controllers, or logic boxes by the OEMs. Similarly, an autopilot is an integral local CPS, while a large CPS could be a collision avoidance system, or a controlled PVC plant system. The global traffic control system includes several independent CPS, several large CPS, and probably some not strictly CPS systems, thus constitutes an ultra large scale system. If some of the CPS comply with the SoS definitions, we may interpret the same hierarchy of CPSs as a SoS, or super-SoS, or ultra large scale SoS (ULSS-SoS). The E.C. projects CPSoS and AMADEOS are good sources about research addressing SoS and CPS.

Note that non-CPS involved within broader systems, could simply be electrical drives, or converters as in the cryogenic example earlier in this section. Studying multiple Cy-Phy boundaries, we may conceptually either abstract the total system into aggregated Cy and Phy parts, or establish a hierarchy of CPS components, sub-CPS, local and global CPS etc. This point may naturally trigger further debates, or lead to philosophical discussions. Figure 1.15 illustrates the situation of multiple Cy, Phy and complete CPS.

In conclusion about the Cy-Phy boundaries: a) broad sense and strict sense boundaries (or microscopic and macroscopic boundaries) depend on the modelling granularity and functional specifications of a CPS as a whole; b) the exact location of the boundary of Cy-Phy may be totally irrelevant, or critical, depending on application details; c) the Cy-Phy boundary may be a crisp line, or a transformational system itself. Calling some small-scale CPS "CPS-Components", may facilitate the combination of the term with other pure physical and electronic systems.

1.5.8 A Simple Classification-Profiling Tool for CPS

Based on the discussion about systems categorisation in Section 1.2.2 we can apply the intuitive scheme proposed in Section 1.2.2.3 to a CPS in a finer version. The resulted combined radar diagram (Figure 1.16) includes for example six dimensions: (1) CPS features/purpose/goals, (2) Socio-technical system orientation, (3) Targeted sectoral orientation, (4) Set of main design

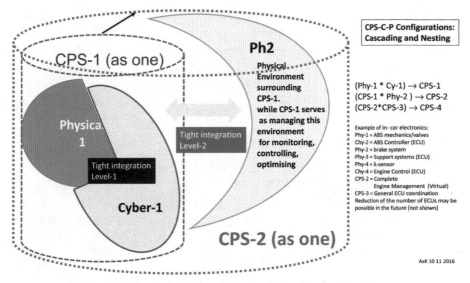

Figure 1.15 Cascading and Nesting Cyber-Physical Systems.

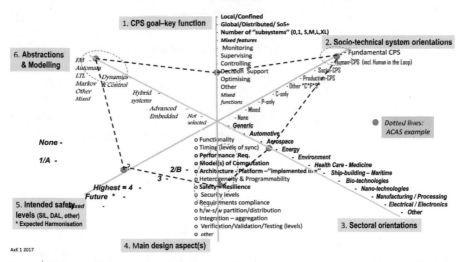

Figure 1.16 Proposed Classification Scheme for CPS. Example mapping of future ACAS-X.

aspect, (5) Intended safety levels and (6) Abstractions and modelling methods. This diagram is envisaged to profile systems, but could be modified to apply to projects.

As an example of CPS classification, the ACAS-X collision avoidance system is shown on the above radar diagram with dotted lines. Here, (1) → Decision Support, (2) → CPS with human in the loop (H-CPS), (3) → sector = Aerospace, (4) → key design aspect = Safety, (5) → intended safety level DAL4 and (6) → rigorous design by formal methods. See Section 1.2.2.3 footnote about DAL (*used in aerospace*).

1.5.9 CPS vs. IoT

At present, a key difference between CPS and IoT is that each sensor in CPS is associated with important local process unit operations, likely also feeds local control loops and therefore there is no collective decision to ignore it. We may call it tight integration of the physical part of the CPS systems as in the CPS definition e.g. X. Koutsoukos and P. Antsaklis [195].

Example: The role of 1,000 temperature sensors in a new large office building for energy monitoring, or 1,000 vibration sensors on a bridge, for structural health monitoring (an example closer to IoT), are very different from 1,000 temperature sensors installed at strategic locations on processing equipment in a large chemical, plastics or metal processing plant (such systems being closer to CPS). But, if the bridge sensors are going to be used for active vibration control, closing several loops and activating suitable dampers, then the initial IoT would need CPS algorithms and coordinated computational sub-systems to achieve the desired goals.

The boundaries between IoT and CPS are frequently blurring. Therefore, industrial applications may relatively easy select the most appropriate concept and architecture for their needs, also combining them into one system.

1.5.10 CPS vs. ICT (Information and Communication Technologies)

In the literature and the sectoral databases/classifications, the new topics have seldom been directly positioned vs. the established since the late 90's "Information & Communication Technologies" (ICT) area, probably because ICT originated from the general computing, communications and office automation fields, not directly linked to the "physical" systems. Furthermore, special industrial computers, process automation/control and embedded systems were not universally accepted as mainstream ICT. Recently, e.g. in a UK survey [150], ICT appears as the superset above all new terms. This is an interesting point, because on one side it may be useful to have a superset on top CPS, however, on the other side, ICT is primarily a set of applied technologies and de-emphasises the early steps such as concepts creation, design methods, tools and system activities involved before products, services and processes reach the markets see EITO [250]. Furthermore, several S&T communities may accept CPS as the ultimate systems class leading to various technologies including large part of what is ICT today. Whether this is a satisfactory position for all stakeholders remains to be confirmed and therefore some more views are necessary to be discussed in the future.

1.5.11 Examples of CPS Challenges and Some R&D Directions

Recommendations for further work, experiences, concrete suggestions and comments about existing gaps have been presented in several papers, reports and projects. Since CPS does not eliminate per se its underpinning topics and technologies, the CPS challenges and R&D directions could broadly be grouped into the following general types:

1. **Pursue core CPS objectives** advocated by its definition but not yet addressed (e.g. truly tight integration of physical and cyber parts in systems), for example by encouraging the CPS community and other relevant scientific, engineering and industrial interested parties to join efforts and fuse domain-knowledge and systems capabilities. We may need novel representations of "physical" laws for direct inclusion into concurrent models aiming at possible straightforward end-to-end solutions.

2. **Discovering totally new CPS relevant ideas** for the development of better and new "systems and control" methods & tools. These may include novel conceptual, mathematical and revolutionary computational approaches (and methods from other fields). There is already a significant effort by several researchers to identify, use and expand, for example co-algebras, novel functional transformations (on signals and dynamics), advanced geometric methods on Graphs & (Semi-) Lattices, high-level "complex numbers/quaternions", regular functions/expressions, multi-scale mixed optimisation & control methods, universal uncertainty modelling via advanced soft sets and more powerful approaches for highly distributed applications in decision making, intelligent agents and cooperative control. Furthermore, Post Hamiltonian, Passivity and dissipativity based designs are also promising ways of modelling complex coupled systems. These are very attractive probably leading to new solutions for classic and emerging challenges.

Examples of such ambitious research directions have been proposed by e.g.: E. Matsikoudis and E. Lee [162, 163], R. Alur [158, 160, 161], J. Sifakis [117, 227], A. Benveniste [122], A. Sangiovanni Vincentelli [121], Patric Cousot [205], P. Antsaklis [182], Moisan et al. [273], Dima et al. [279].

3. **Continuation and improvements of the underpinning approaches in CPS sub-domains**, initiated at earlier times under e.g. embedded systems and control. Emphasis must be placed on the enhancements and combinations of already known and very promising methods, languages and tools (for example BIP, MIC, Contract and Platform-based designs; tools such as UppAal, Astree), to overcome current limitations of today's methods and become integrated design environments for industry. High Confidence Run-Time Verification [120, 210, 246] and robust CPS security [132] are just two examples of well focused R&D. Compared to the past, the extra push today, also comes from the interdisciplinary holistic orientation and encouragement of the CPS concept itself, combined with emerging powerful new hardware and control approaches [251, 256, 257]. The Mixed Criticality Systems domain (MCS) is another attractive R&D field with e.g. scalability, particioning and certification challenges [95–99].

4. **Applications**: Since CPS is already key part of several areas in industry, services and societal applications, with, or without humans in the Loop, its application will be possible without much new effort. Almost all sectors have embraced CPS: "CPS for transportation, energy and environment", "CPS for Healthcare", "CPS in security and security of CPS", "CPS in Robotics and Telemedicine", "CPS in IoT/I2oT", "CPS in defence". But other CPS opportunities, less reported today, arise e.g. in nanotechnologies and nanomaterial production, marine & ocean engineering, fusion energy, precision agriculture, arts, sports, hobbies and music.

5. **Combined CPS with e.g. SoS, IoT, HPC, 5G, cloud, big data and analytics**, which are briefly discussed in other parts of this chapter. See also the CPSoS and AMADEOS project reports. Utilisation of generic (cross-sectoral) CPS developments and earlier methods and tools in the other new fields, for example CPS methods and tools can help realise safe and secure Industrial Internet of Things (I2oT), similarly to achieve extreme low latency systems required for the design and implementation of the 5th Generation mobile environments.

Additional systems R&D themes, beyond CPS, are given in Section 1.10.3 of
this chapter.

1.6 IoT in More Detail and the 5G Mobile Technologies

1.6.1 IoT Applications and Benefits – Internet of Everything

As indicated in Section 1.4.1, IoT started as extension of RFIDs and WSNs
and evolved as a new networking technological opportunity different from
previous fixed architectures. It initially aimed at applications where large
number of sensors (usually low cost and frequently in redundant configurations) and some output activation devices, or normal actuators are (wirelessly)
linked to appliances, within buildings, or infrastructures and via lightweight
integration platforms. These achieve, or contribute to global goals, for example monitoring, situational assessment and local non-critical controls and
in this way the integrated IoT presents an overall "system status" to the
responsible agency. IoT offers indeed new ways of collecting massive sensors
data, engage software controllers and activate networked actuators, making
them available for processing and decision making, under novel architectures
and configurations (such as ad-hoc, self-*, cloud, edge and fog-based). IoT
solutions are easier to implement, than dedicated/wired ones, and therefore
suitable, for example, for innovative monitoring applications if they meet the
required safety and resilience requirements of the case.

The economic advantages, ease of installation and programming of such
configurations of sensing and decision making capabilities, offer not only new
solutions to some existing engineering challenges, but also became obviously
attractive in addressing new applications. For example, the so called Internet-
of-Trees (e.g. as tested at Harvard campus) and health monitoring of large
civil constructions (buildings tunnels, bridges). This expandability potential
of the initial IoT concept and technology led recently to the conclusion,
within the supporting organisations and the S&T communities, that *IoT and
Internet of everything can do much more than wireless sensing for certain
applications.*

1.6.2 On the IoT Definition

The IEEE society for IoT has prepared a study[13], where both the IoT evolution and the various definitions proposed before, are discussed in depth

---

[13]http://iot.ieee.org/images/files/pdf/IEEE_IoT_Towards_Definition_Internet_of_Things_
Issue1_14MAY15.pdf

and the clear and unclear points of the previous definitions highlighted and analysed. The report also presents the views of this group concerning the scope of CPS and concludes that "CPS is a closed system, in contrast to IoT". This IEEE group proposes a two-level definition based on the scale of the IoT systems:

IoT for low complexity systems scenario: An IoT is a network that connects uniquely identifiable Things to the Internet. The *"Things"* have sensing/actuation and potential programmability capabilities. They can be collected and the state of the "Thing" can be changed from anywhere, anytime, by anything.

IoT for large environment scenario: Internet of Things envisions a self-configuring, adaptive, complex network that interconnects "things" to the Internet using standard communication protocols. The interconnected things have physical, or virtual representation in the digital world, sensing/actuation capability, a programmability feature and are uniquely identifiable.

Note that the European Research Cluster, IERC, (a European Union-funded project aimed at addressing the large potential for IoT-based capabilities in Europe and to coordinate the convergence of ongoing activities) states that IoT is: *"A dynamic global network infrastructure with self-configuring capabilities based on standard and interoperable communication protocols where physical and virtual "things" have identities, physical attributes and virtual personalities and use intelligent interfaces, and are seamlessly integrated into the information network".*

1.6.3 Industry Views on IoT and the Industrial IoT (I2oT) by the I2 Consortium

While the above-mentioned IEEE IoT study does not discuss the IoT security issues, it is worth noting that in 2014, a very strong international "Industrial Internet consortium" called I2C was established to develop an Internet of Things with strong security specifications. Few years earlier industry did not endorse IoT protocols such as ZigBee[14]. Even the newer professional version of ZigBee was short of expectations regarding the encryption and its security requirements. Up to now two protocols have been used in industry (approved

---

[14]Tomas Lennvall, Stefan Svensson and Fredrik Hekland: A Comparison of Wireless-HART and ZigBee for Industrial Applications, http://cwi.unik.no/images/A_Comparison_of_WirelessHART_and_ZigBee_for_Industrial_Applications.pdf

by IEC/standard 62591), namely Wireless HART[15] and ISA100 [252], the latter is also IPv6 compatible. Furthermore, the German Industry 4.0 Initiative is already cooperating with the Industrial Internet consortium.

1.6.4 5th Generation Mobile Technologies (Expected around 2020)

The 2nd generation mobile technology (2G) was about global voice; 3G was about voice and data; 4G was about voice, data and applications[16]. Today, supported by many S&T organisations, major equipment manufacturers, telecommunication operators, researchers and business stockholders, 5G is an advanced, higher carrier frequencies (likely to be above 6 GHz wi-fi and 300 GHz links) broadband multi-technological topic, encompassing cellular and satellite solutions and aiming at deployment around 2020. It refers to a significant expansion of the previous generations of mobile technologies, in terms of capacity and resilience, and it will be a key enabler for the Internet of Things and I2oT, *"by providing interactions with a massive number of sensors, rendering devices and actuators with stringent energy and transmission constraints"* [185].

However, to implement the 5th generation networks, it requires several new S&T developments already underway, such as a) improved modulation and coding techniques, i.e. more efficient than today's Orthogonal Frequency Division multiplexing (OFDMA) and signal processing by e.g. short-term Fourier transform (STFT) and filter bank multi-carrier (FBMC), b) denser cellular networks, which would require suitable backbone and new core and radio access network (RAN) layers, probably based not only on fibre, but also millimetre wave wireless themselves in combination with satellite services, c) in turn, this means suitable beam-forming type antennas, for advanced space division multiple access (+SDMA), d) use of network cloudification, e) handover coordination between mission critical heterogeneous air interfaces e.g. M2M applications, 3D telepresence on mobile devices and the important requirement of reduced end-to-end latencies of the order of 5ms are needed to support interactive applications. An ambitious goal is to ensure latencies of $\leq$1ms.

---

[15]http://processonline.com.au/content/wireless/article/applying-wireless-to-ethernet-ip-automation-systems-part-2-1332415582#ixzz4T7VHnbvh

[16]Alexander Hellemans, Why IoT Needs 5G, IEEE Spectrum, May 2015.

*"The deployment of ultra-dense mobile networks with numerous small cells will require new interference mitigation, backhauling and installation techniques. 5G will change the role data is playing across our built environment, for day to day asset monitoring, and maintenance decision-making. 5G represents a new turning in how to design communication systems, taking a holistic approach to connectivity, as a **network of networks**; the big picture question is how everything will fit together with each other. In particular, peak data rates will be in the order of 10 Gb/s and a capacity of 10 Tb/s/km^2 will be required to cover e.g. a stadium with 30.000 devices relaying the event in social networks at 50 Mb/s"* [185–190].

Comment: While many of the challenging topics in 5G may benefit from advances in CPS (e.g. MIMO Control, DEVS and Power management optimisation tools) and from SoS (e.g. large scale architectures, emergence management and cloud/edge-SoA applications), the initial 5G developments seem to reside within the telecommunications disciplines, industries and supporting academic institutions. Therefore, a strong cross-disciplinary cooperation would help for example in resolving quality of services, achieving practically "near zero latencies", high product dependability and long tern evolvability issues in the ambitious 5G endeavours.

1.7 System of Systems (SoS) and SoS Engineering (SoSE) – More Details

M. W. Maier (1998) [91] defined five key characteristics for a SoS: (1) Operational independence of component systems, (2) Managerial independence of component systems, (3) Geographical distribution, (4) Emergent behaviour and (5) Evolutionary development processes. A useful definition, out of several ones, is proposed by Mo Jamshidi, 2009 [17] *"SoS is the integration of a finite number of constituent systems which are independent, able to operate and which are networked together for a period of time to achieve a certain higher goal"*. However, frequently the systems attempted to be organised by a SoS engineering method, may only partially comply with all 5 criteria. A typical exception is the extent of the *geographic distribution*. While it is a matter of convention, a basic minimal requirement for a strict-sense SoS could be the combination of {(*1*) & (*2*) and (*4*) & (*5*)}. DeLaurentis and Crossley [92] have added to the five SoS characteristics mentioned above, few more, namely: *"inter-disciplinarity, heterogeneity of the systems involved, and networks of systems"*. The IoT with IPv6 and the forthcoming Industrial Internet of Things (I2oT) with a 5G architecture may be of great interest in such Networked SoS. A comprehensive report on the state of the art of SoS/E is conducted by the EU-US project T-AREA-SoS and can be found in [191].

Another interesting aspects of a SoS is its activation period. In the "Guide to the Systems Engineering Body of Knowledge (SEBoK)" [89], it is stated that, *"the formation of a SoS is not necessarily a permanent phenomenon, but rather a matter of necessity for integrating and networking systems in a coordinated way for specific goals such as robustness, cost, efficiency, etc"*. Nevertheless, we consider that the activation of the overarching SoS structure and underpinning ICT infrastructure need to be available at any time on demand, to gain the benefits of the overall SoS. Transitional delaying phenomena to set up SoS may be against the effectiveness of the concept [93].

1.7.1 System of Systems Examples

The first large European Commission (EC) funded projects in SoS/E e.g. DANSE, COMPASS, Road2SoS, T-AREA SoS, CPSoS, AMADEOS [292]) and also INCOSE reports and SEBok [89] provide an excellent set of SoS/E applications. To give an elementary example presenting SoS/E as a candidate solution, we elaborate a typical City public transportation services case. Figure 1.17 illustrates a possible System of Systems view of a big city Bus Transportation and Citizen Information Service in simplified form. It meets almost all Maier criteria namely: managerial and organisational independence, *medium level geography spread*, several emergent and long term evolutionary behaviours in everyday operation and lifecycle. Such a system is more complicated than the illustration, because it normally addresses also e.g. advanced messaging and internet services and intermodal connections, urban and regional aspects of transportation, such as city connections to hubs for airports, ports and train stations for other destinations. In such an extended case, the *"geographic distribution"* criterion is then fully applied.

Note: In the above City SoS example, some of the constituent systems may be CPS-driven (e.g. requiring time synchronisation, optimisation), designed according to CPS methods (electric, or other buses), or use IoT's (Telematics, mobile phones services traffic, surveillance cameras). The centre of gravity of

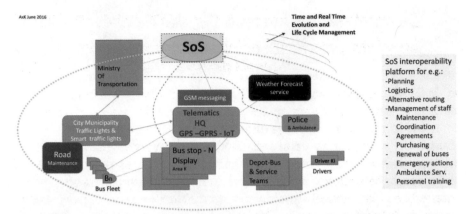

Figure 1.17   Example of a simplified SoS and its constituent systems related to a typical Bus Transportation and Citizen Information Service.

the overall SoS is in general on organisation, coordination and management while all its constituent systems demonstrate their strong CPS nature.

1.7.2 Dahmann and Baldwin Types of SoS

Dahmann and Baldwin [90], defined four types of SoS summarised here: "*Directed* – the constituent systems are subordinated to the SoS. The component systems maintain an ability to operate independently; however, their normal operational mode is under a central managed purpose; *Acknowledged* – The SoS has recognized objectives and changes in the systems are based on cooperative agreements between the SoS and the system; *Collaborative* – The component systems interact voluntarily to fulfil agreed upon central purposes. The central players collectively decide how to provide, or deny service, thereby providing some means of enforcing and maintaining standards; and *Virtual* – The SoS lacks a central management authority and a centrally agreed upon purpose for the SoS. Large-scale behaviour emergesand this may be desirable, "but this type of SoS must rely on relatively invisible mechanisms to maintain it." [89]. Figure 1.18, based on [90], is an attempt to schematically interpret these four SoS types.

Comment: If an overarching architecture and ICT system is developed and applied, this SoS would exhibit collaborative and few virtual characteristics as schematically interpreted in Figure 1.18(b). Few Indicative (but not confirmed) examples of possible organisations and companies that could

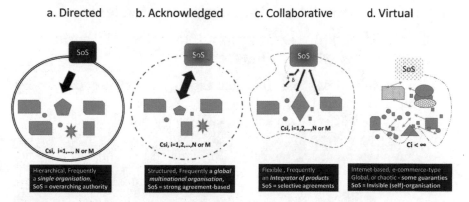

a. Directed   b. Acknowledged   c. Collaborative   d. Virtual

Figure 1.18   Main SoS categoried according to the DoD Guide (2008) and Dahmann & Baldwin.

belong to the category indicated may be: Directed (Armed Forces, Government services), Acknowledged (HP, Shell, Unilever), Collaborative (Olympus Cameras, Daimler-Chrysler, Airbus, Ferrari cars) and Virtual (Amazon, e-Bay, London Taxis, Uber), but it is not implied, that such organisations actually use SoSE methods, or are structured accordingly.

1.7.3 Large Scale Systems, Complexity and (Old and New) Cybernetics

The theoretical R&D on complex systems and control is at least 60 years old and continuous today to be useful towards new challenges such as modelling of emergent phenomena, passivity based control, fault tolerance and dissipative systems [182, 194]. N. Wiener' work on Cybernetics was oriented towards signals and systems, Cybernetics today became very broad and includes a significant community from philosophy, through systems to engineering. Furthermore, New Cybernetics, or 2nd Order Cybernetics, is a branch of today's systems and complexity R&D community[17] beyond SoS/SoSE.

Note: Several large scale CPS are not always declared as SoS, although they may exhibit SoS characteristics.

---

[17]See A. Fradkov [19, 201], Novikov [154] and Nick Karkanias, Workshop on: "Complex Systems and Control", 21–22 July 2016, City University, London, UK.

Part II – Decision Making, Processes and Systems

1.8 Decision Making: Definitions, Examples, Methods, Interactions with System Design

Decision Making (DM) and Decision Processes (DeP, or DP) are at the heart of any goal-driven activity, either pure human and social, or within technological and computer-based systems, or in combination of both. It may be a very small part in the logic of a system, for example a shut-down process termination (in hardware, or software, but not easily recognisable as a separate entity), or a complicated, large evaluation process with many factors and criteria, leading to specific conclusions, selection of optimal/sub-optimal values and finally triggering further actions. In brief, DM covers technological and non-technological applications and can be implemented in automated, semi-automated, or manual modes. Recent debates among industry leaders, declared that "The biggest risk for companies nowadays can be the way they make decisions" [199].

1.8.1 Scientific, Engineering Aspects – Machine and Human Decision Making (DM)

DM is tightly coupled with systems and control. In certain situations, a DM problem appears in the literature and in practice as a system design challenge. However, the scope of DM is very broad and it constitutes a separate discipline with many stakeholders and rich underpinning scientific repertoire. DM is essentially a problem-solving approach, or activity of humans and machines (e.g. agents), see [11, 14]. DM is also considered as part of cognitive and social sciences, but also contributes to a broader landscape, spanning from psychology, economics, philosophy, risk management, engineering and several branches of mathematical logic and complexity. DM is an interdisciplinary field. Furthermore, in DM, some special concepts and terms are used, not traditionally found in systems sciences and engineering, such as *"preferences"*, *"beliefs"*, *"values"*, *"prospects"* and alternative views related to uncertainty, for example *"personal/subjective judgment"*, *"ambiguities"* and *"ambivalent"* situations e.g. [260].

Table 1.3 presents several scientific DM methods with examples of other fields using them.

Table 1.3 Scientific Methods used in modelling Decision Processes for Decision Making and building Decision Support Systems

Topic	Modelling and Implementation using:	Same method used in other areas. Examples: ↓	Notes
Decision Making (DM)	Intelligent Agents (e.g. supervised) Optimisation, or other approaches (e.g. Utility functions, Expected Utility values)	Communications, Cooperative systems & Control, Planning, Applications in CPS, Internet of Things, Social Networks,	Basic Model of DM: (observation-valuation-selection criteria-action). Usually multiple agents
	Probabilistic Bayesian theories Random Graphs, Trees	Stochastic Estimation, Control, Optimisation, CPS, IoT	
	Dynamic Programming (DP) Hybrid Systems, Graphs, MI-LP/NLP, Variational Calculus	Control,, Scheduling, CPS, Optimisation, Planning, Scheduling e.g. RTOS	DP to support MDP Verification through other methods
	Reachability Theories, Soft Sets	Control, Cooperative Systems	Scalability for large applications
	Markov decision Processes (MDP)	Operations Research, SoS/E	
	Partially Observable Markov Decision Process Distributed/Decentralised POMDP	Strategic Choices, Collision Avoidance Modelling. Sequential processes Management Social Networks Planning	Verification through other methods, Scalability challenges if many distributed entities involved
	Games Theories e.g. DG, POSG	Economics, Options,	Useful Enhancements exist
	Learning Theories (e.g. Supervised, Reinforced)	e.g. AI, Control, Robotics	
	Other methods, New concepts		Quantum DM Genetic Algorithms

1.8.2 Definitions and Basic Considerations

A simplistic definition from the web is: *"Decision Making (DM) is when we conclude discussions and start actions"*. A more elaborate definition is: Decision making is the process of selecting (if possible) a preferred/favourable option O_k out of several available ones, based on specific criteria, beliefs or preferences. $(O_1, O_2, \ldots, O_N) \rightarrow O_k, (k \leq N)$ to achieve a goal. An assumption, concerning human-based decision is the rational-reasonable human behaviour, which however is not always valid.

Concerning computer-based systems and based on M. Kochenderfer [14], DM is defined through Intelligent agents which are "general entities (physical such as humans, or robots, or non-physical such as decision support systems). Then, DM is the process of an agent registering observations from its environment $(Ot_1, Ot_2, \ldots, Ot_N)$ (Note: either received at discrete times $t_1, t_2, t_3, \ldots, t_k$ (last one at time t_k), or are available all together at once). The agents must evaluate the implications of their actions, through a set of criteria, and chose an action A_k best meeting the goal(s) of the DM process".

Figure 1.19 illustrates the general decision process in a feedback loop leading to iterations to reach a final choice. In this pictorial diagram we

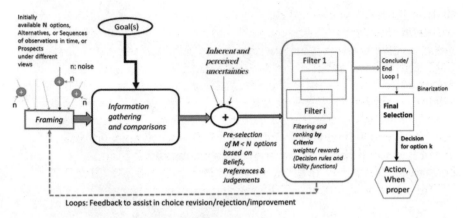

Figure 1.19   Example of a system model illustrating a general decision making process.

inserted a process "termination block" to avoid endless attempts at finding the optimum selection. This may not always be an easy step and the question emerges *"Can every decision problem posed converge to a crisp decision?"*

Collective decision making

There are some approaches to deal with complex decision problems, or resolve problems of not reaching decisions, for example in cases of complicated, or extremely uncertain conditions, or under clear conditions, but when conflicts exist between set goals, given system models, selection criteria, available options and the allowable time-window to conclude the process. The situation is like the elementary control valve decidability example discussed in Section 1.8.4.2.

In human-driven collective decisions, that is, having many *"experts"* (e.g. $N = 2n + 1$), trying to decide about the selection among certain options (m), where n and m are integers, there are practical approaches used in industry, for example:

Automated unsupervised majority voting, which however may swiftly conclude to a decision, but which may lead to the wrong choice due to ill-conditioned parametrisation. Another one is supervised consensus building with the help of a moderator, who, in case of a deadlock between opinions, will use new arguments and explain the agreed criteria, or ultimately add few more experts ($2n + 1 \rightarrow 2(n + k) + 1$) and repeat the process. Usually, relaxing, or changing the criteria/rules is not fair, nor acceptable for a given goal of the process. The process does not guarantee convergence and the moderator may resort to majority voting.

In machine DM, a) in case of small systems, or networks, a type of majority voting like the 2 out of 3 logic, or similar may be sufficient to implement through intelligent agents assigned for the DM and b) in large-scale systems with many DM alternatives in distributed environments, the problem complexity grows, leading to difficulties like the ones encountered in distributed optimisation [253–255]. In such DM cases, distributed agents will need larger amounts of information and an overall DM coordination with additional rules.

Examples of DM models and underpinning theories: 1) Utility theories and functions for assigning numerical values to ranking and preferences; 2) Markov decision processes (MDP) and dynamic programming (primarily for modelling and optimising sequential processes); 3) special Markov processes (modelling multiple agents in cooperative DM through decentralised MDP and Partially Observable Markov Decision processes (POMDP); 4) advanced distributed Bayesian networks and decision graphs for Gaussian and other distributions, such as Gaussian sums; 5) behavioural games theories; 6) Quantum and Adaptive decision making, 7) concerning human and social DM, high level meta-cognition theories and 8) Actors Network Theories (ANT) and Agency theories and 9) multiple objectives DM. These topics are discussed for example in [258–270, 272, 278, 283, 293–295].

Binarization and Undecidability

The term binarization is used e.g. in image processing, meaning the conversion of a colour image to a black & white one. It is a reduction process preserving important characteristics of the original finite image by mapping certain pixels to binary choices i.e. $(x_1, x_2, x_3, \ldots, x_n) \rightarrow (0, 1)$. This may not be at least theoretically possible in all other cases though. In general, this problem belongs to the so called *"Decidability Problems"* where we investigate the existence of a [*yes-* or *no*] answer to an infinite set of inputs, with certain (only) inputs returning a *yes* answer[18,19] [218].

Concerning DM within a CPS system-thinking framework, Lee & Seshia presented and analysed in detail the problem of conversion of information to

---

[18]https://en.wikipedia.org/wiki/Decision_problem
[19]https://en.wikipedia.org/wiki/Thresholding_(image_processing)

a [*yes, no*] decision as a Turing non-computability and undecidability challenge [11]. Also R. Alur discusses the non-computability problems related to the (automated) verification challenge when applied to systems with too many states and transitions, as a computational problem [12]. Stavros Tripakis [203] presents similar challenges concerning the difficulties to synthesise decentralised controllers for communication protocols. M. Kochenderfer [14] talks about undecidability of the infinite horizon Decentralised Partially Observable Markov Decision Process (DPOMP), difficulty which is practically due to the complexity of the huge amount of options each decision agent has to evaluate. Finally, Hu and Kaneko discuss the positive, negative decisions and undecidability related to two person games [202] which may be useful in CPS DM applications.

1.8.3 Motivating DM Example: Football Goal Line Technology

A well-known example of decision making is emerging in sports, e.g. in football, the so-called line pass technology, compared to human only judgement [281, 282]. The following Figure 1.20 illustrates two technical DM solutions based on different measurement techniques, (optical/vision and proximity detection) and two human-based approaches, one only expert and the collective decision of two experts. The core challenge in the technical systems involved is the decision about "*where to set the threshold(s)*" of identified

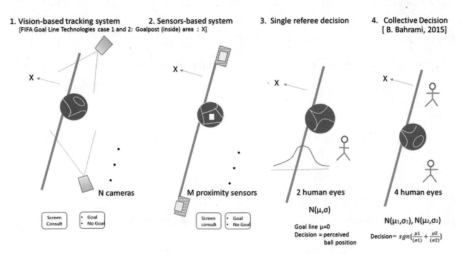

Figure 1.20 Four examples of decision making approaches for declaring a ball in the goalposts (Automated, Machine-Advisory, Human).

ball-violation of the line. Assuming rational human decision making there are advantages of collective DM. Case 4 in *Figure 1.20 is based on: B Bahrami et al. "What failure in collective decision-making tells us about metacognition, Phil. Trans. R. Soc. B (2012) 367, 1350–1365. doi:10.1098/rstb.2011.0420,* http://rstb.royalsocietypublishing.org/content/367/1594/1350, and on collective decision making from a psychology point of view: *Bahrami, B., Olsen, K., Bang, D., Roepstorff, A., Rees, G., & Frith, C. (2012). Together, slowly but surely: The role of social interaction and feedback on the build-up of benefit in collective decision-making. Journal of Experimental Psychology: Human Perception and Performance, 38(1), 3–8. doi:10.1037/a0025708.*

1.8.4 Industrial Examples

1.8.4.1 Same decision making challenges in different industries

Example: Material Quality Control, Automatic Decision Making and final Action(s).

On-line material processing and real-time quality control is a major DM task in several industries (Real-Time Decision Support systems (RT-DSS)). For example: a) to visually inspect wood slices and automatically cut out low quality parts, b) similarly in steel making (cold, or hot mill production) to mark-up acceptable sections and identify reject quality, c) in the textile industry concerning fabric quality and d) in pulp and paper industry. The solution in all these cases is a real-time decision logic to control the separation/classification/discrimination process. At earlier times this was done by human operators visually inspecting products with optical tools, using comparisons against a set of known patterns. Advanced computer vision systems (when accuracy surpassed multiple human eyes capabilities) were later introduced. The DM logic automatically energises cutting, or marking actuators to classify the good and the other qualities. Hardware (e.g. FPGAs), or software implemented logic were used depending on several factors such as speed of moving surfaces, reflectance, type of colour temperature, types of erroneous surfaces, in relation to the accuracy of the intelligent system.

Comment: The design of the above systems as CPS or as DSS, may in principle be the same, but in practice, may also be different because of possibly different dominant system architecture factors and required characteristics of the material under an overall quality control system with a DM subsystem. The performance of any solution would depend on *functional* and *non-functional* requirements of the overall system. See Section 1.9.

Note: These quality control systems are intended to operate automatically without human in the loop, except mainly in break-down and maintenance situations. A critical DeP stage is the selection of *thresholds and boundaries* of acceptable and non-acceptable material.

1.8.4.2 Elementary process automation DM – Decidability cases

Example: Controlling a "critical", with limited physical access, *remote on-off valve* in industry, seems to be a very simple engineering task for a digital software-based computer control system (see Figure 1.21). Let us study this case and assume, that at time (t_1-) the last valve state was (electrically energized) and actually open (V=open). Critical means here that abnormal operation may have high economic and/or operational integrity or safety implications. This simple decision process looks like:

> High level software instruction in real time: "V=close" at time t, or upon an event.

> This changes/sets a bit in the digital output for this valve DO1-V.

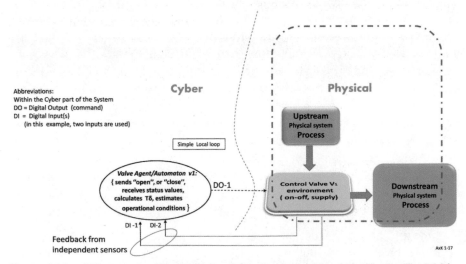

Figure 1.21   CPS example of a simple, local, low level actuator-monitor logic. The DM in local CPS. This system is part of a larger CPS.

But to proceed, one may ask: *"how does the system know that the valve has really closed?"* In practice, indeed we use more elaborate considerations to ensure better design and correct execution in real time:

The valve may be equipped with two limit switches S_h, S_l, associated to two digital inputs S_h, S_l (DI is a binary variable). In the initial state of the valve (open) we assume that the upper switch is closed and the bottom switch open, we also assume, that the normal travel time Δ of the stem of the valve is T_N. We use $\wedge$ for AND logic, ◆ for eventually true and U for until; $\oplus$ is an exclusive OR gate to deduct inconsistencies.

The automatic system can additionally trigger a timer to calculate the actual time T, estimated from the time-stamped status of the two switches. The decision to declare, that the valve has really closed, can be based on:

- *one criterion*: for example that the bottom switch (S_h) is closed, as sensed by its DI
- *two criteria*: ◆ (top switch S_h opened) $\wedge$ ◆ (bottom switch Sl closed))
- *three criteria*: $((S_h$ open) $\wedge$ $(U$ Si closed)) $\wedge$ $(0 < \text{T} \leq T_N)$

When conditions are not met (various combinations), alarms and other actions will be triggered by the decision logic. For example wrong $(S_h \oplus S_l)$ implies faulty switches. Additional inferential measurements may also be used, based not only on the combinatorial status of the above variables (S_h, S_l, Δ), but on appropriate process variables related to the implications of the operation of this valve on the rest of the process.

Undecidability cases may arise, (beyond common mode failures in the channels), such as double faults in the switches, or certain malfunctions (of the computation) masking the real situation.

The designer of DM systems will also consider non-functional requirements of the real-time software system implementing the logic. This example emphasises the complexity of the automated decision logic for such an elementary case. Figure 1.21 illustrates the valve logic evaluated by an agent (or a deterministic automaton) for the above analysis.

Note: A deterministic finite automaton (DFA) is a tuple Π: $(S, \Sigma, \delta, S0, F)$, where S is the set of states, Σ the set of triggers, δ is a function $S \times \Sigma \rightarrow S$ denoting the transitions from state to state, $S0$ is the initial state of the automaton and F is a set of final states (could be one) [284–286]. A DFA resembles a discrete control loop/system and serves to model/describe it. The programming of industrial PLCs and other control systems have extensively been supported by DFA methods and tools.

1.8.4.3 Decision making and processes in large scale Collision Avoidance systems (CA)

We discuss further here the introductory example, about the forthcoming Mid-air Collision Avoidance (CA) system ACAS-X, which can be interpreted through several different views including decision making. Similar challenges are encountered in other practical industrial applications, summarized in Table 1.4 which demonstrate the high industrial and social interest in managing tracking objects and CA systems. Note that Figure 1.13 in Section 1.5.5 illustrated the multiple cranes application as a CPS challenge, but part of the coordination and safety logic includes DM processes (e.g. criteria for "abort" and "suspend" states).

ACAS-X Collision Avoidance System as Decision Making challenge

The new international ACAS-X collision avoidance (CA) system concept, being tested and refined since 2013 was briefly introduced in Section 1.1.2 of this chapter as part of future Air Traffic Control (ATC) systems (under Next Gen in US and SESAR in Europe). Furthermore, in Section 1.5.5 it was commented that ATC (including ACAS-X) could be conceived and modelled as either large distributed CPS (dynamics and control), SoS (coordination

Table 1.4 Similar technological challenges in different domains and sectors. Usually different solutions and standards apply

Compiled by AxK 11-2016

Collision Avoidance in:	Current needs for improved Systems :	Contributions by CPS, IoT, SoS/E, Decision Making (DM), Other
Aerospace: • ATC /ATM	- Mid-air aircraft collision avoidance (CA), - Civil-Military and UAVs co-existence - Geo-fencing considerations, GPS - On the ground CA (Taxi, Runway and Ramp)	{CPS, Control, SoSE and DM; On the ground also {IoT sensing, Communication Systems, Cloud}
Civil Engineering: • Construction sites • Ports Loading Terminals Industrial Engineering	- Multiple Cranes ground operations - Multiple loading cranes on-board ships - AGVs movements on the ground, GPS - Human–Machines safe & secure interactions	{ CPS, DM} Ports organisation: also{ SoSE + Communication Systems } {Robotics, I2oT, M2M, SCADA, MES}
Land Transportation: • Rail • Automotive/Highways • City systems • Pedestrian movements Sea Transportation	- Monitoring, Control, Level-crossing protection - Autonomous Driving and related designs - Traffic lights control, Accidents prevention - Interaction at road crossings/ restricted areas - Autonomous sailing. Proximity to ports/Land	{ CPS, Control, IoT, SoSE, DM, Communications, Antennas, Cloud, Edge, Fog, 5G, I2oT, Cognition, Vision based systems} GPS and similar systems
Sports, Music, Stage events	- Slightly different situations such as Line-Pass Technologies, Robotic Props movements	{CPS, Control, Vision, Robotics, IoT}
Toys , Educational set-ups	- N/A, but faster decisions & control desirable	CPS, IoT and DM, vision tracking,GPS

and quality), or DM systems (decision support). In fact this CA problem was first formulated in mid-90's using rigorous hybrid dynamic systems models [48], (topics actually in CPS today). In more recent years, since 2009, the R&D efforts guided by the aviation standardisation authorities, targeted on top of the a/c dynamics also the practical implementation of such an on-board system. The current ACAS-X can therefore be considered as a combination of several intelligent subsystems mapped into embedded (CPS) software modules, which will be integrated into a coherent overall system. One of its modules is a decision support algorithm and database (DSS) which would be used to advise the pilots to avoid risky trajectories. A simplified diagram, based on an ACAS-X system interpretation is shown in Figure 1.22 [57]. It includes off-line and real-time parts. While the individual modules and their exact algorithms are not yet finalised/published, it is worth mentioning, that the fundamental method to meet the first FAA requirements was initially a dynamic programming optimisation of comprehensive off-line flight data, combined with real-time sensor-based updating of the current position of an a/c and potential "intruder a/c" in its vicinity. This system uses a smart lookup table and decision rules to automatically formulate and issue advisories to the pilots. See M. Kochenderfer, J. Holand and J. Chryssanthakopoulos [51]. But the "last mile" of these ambitious efforts, although close to a real-life

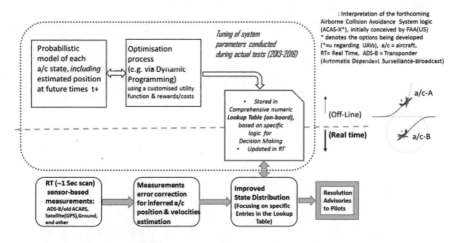

Sources: Diagram based on Eurocontrol, Netalert Newsletter, p. 1-4, June 2013

Figure 1.22 Interpretation of the forthcoming Airborne Collision Avoidance System logic ACAS-X. (X is a qualifier of special classes, e.g. U adressing UAVs).

application, is still ahead us (2016) before manufacturing, certification and deployment [52–56].

IoT-driven large scale systems DM applications

Concerning applications of large scale systems with emphasis on IoT – the software coordination platform and related communication protocols are important elements to be considered. [221] presents an interesting detailed study related to two EC funded projects Local4Global and HARPA. In this R&D work the multi-aspect considerations by all three types of systems i.e. CPS, IoT and SoS prove beneficial for system integration and provide us with another example of potential synergies among the new topics.

Comment: As introduced in Section 1.1 of this chapter and briefly discussed in Section 1.5.5 under a CPS thinking, the Collision Avoidance (CA) topic has attracted high interest among the aeronautics, systems engineering and control research communities, who recently provided insightful analysis, formal validations of the algorithms and also proposed alternative approaches for its modelling and design. Feasibility and performance of the DM methods were also validated after realistic flight conditions by A. Platzer et al. [54, 204, 206]. The topic of dynamically estimating critical mutual positions of two, or more aircrafts (a/c) in a given 3-D space with several uncertainties involved, was introduced in Section 1.1. In cases of more than two aircrafts, the complexity increases, but smart methods have also been proposed such as in [61]. Furthermore, examples of large scale systems with a clear IoT flavour (e.g. as in [221]) show the emphasis-shift towards the overall system management, than on the more CPS-type optimisation of the DM approach at local level (node level). In both cases though the design of the exit logic is important.

In conclusion, the key objectives of collision avoidance in airspace are: avoid critical conflicts by issuing easy to understand alert to pilots, while minimising the frequency of alerts for non-critical encounters (in process automation jargon this would mean "trust the alarm system and supress nuisance alarms"). Beyond aerospace, other industries use specific guidelines to reduce unnecessary alerts[20]. From a S&T point of view, more involvement of the CPS and SoSE communities in all these developments seems to be appropriate.

---

[20] see e.g. Engineering Equipment and Materials Users Association, EEMUA Guide No 191 and BS EN 62682:2015 Management of alarms systems for the process industries.

1.8.4.4 Consensus based methods, majority and supermajority voting

Following the remarks in Section 1.8.2, consensus decision-making is a group decision-making process in which group members develop, and agree to support a decision in the best interest of the whole, therefore the DM process is a kind of optimisation process[21].

Since for a general set of utility values $x_i \to \min x_i \leq \frac{\sum(\lambda_i \times x_i)}{\sum \lambda_i} \leq \max x_i$, implies that human based consensus is frequently a compromise. For example, beyond the standard industrial solution of 2:3 majority voting in redundant, but more expensive configurations, (we need to provide 3 measurements for the same variable), more advanced approaches have been applied such as "qualified majority as a judgment-aggregation rule which assigns, to each profile, the collective judgment, considering some acceptance threshold". This is called supermajority rule when the acceptance threshold requires more than a simple majority of the involved experts ("50+k%"). A less radical approach would be that agents have bands-of-belief as well as binary (all or-nothing) beliefs".

Furthermore, in cooperative control applications "The consensus problem is to have a group of vehicles (or more general agents) reach a common assessment, or decision based on distributed information and a communications protocol" [107]. Using agents, a basic consensus algorithm is modelled as $\frac{d(x_i)}{dt} = \Sigma \lambda_{ij}(t)(x_j - x_i)$, where x_i are the information states of vehicle i and $\lambda_{ij}(t)$ reflect the degree of communication between vehicle i and j. See also [111, 207].

1.8.4.5 Human in the loop (HitL)

Example 1: Large scale Agricultural Automatic Inspection and human decision: *"to spray, or not spray."*

An interesting DM application reported[22] uses remote sensing, and UAV-based imaging of large fields, to automatically evaluate and classify the state and quality of crop cultivation areas per geo-location, followed by human in the loop final decision to spray, or not during a specific time of the year. This is a non-safety critical Human in the Loop application[23].

---

[21] http://wikipedia.org/wiki/Consensus_decision-making

[22] Terry M. Stewart and Mark E. Brown: Proceedings Ascilite Auckland 2009, http://www.ascilite.org/conferences/auckland09/procs/stewart.pdf

[23] See also Emilio Gil: Advanced Technologies for the Improvement of Spray Application Techniques in Spanish Viticulture: An Overview, Sensors 2014, 14, 691–708; doi:10.3390/s140100691

Example 2: Decision to switch from automatic to manual operation in future Fusion reactors gas processing systems (simplified scenario).

Future reactors for fusion energy include a gas processing plant to purify and recycle fusion fuel (Hydrogen isotopes, in short H/D/T)[24]. The process is very interesting from instrumentation and CPS control point of view, but goes beyond the scope of this chapter. Only a small DM case will be briefly presented here. The purification process is intended to operate automatically based on a Petri Net logic concerning expected composition of incoming gases with some real-time corrections based on in-situ analytic measurements. However, if the measurements are out of expected ranges, suitable automatic execution of established processing sequences will be impossible to apply and therefore (in order not to shut down the complete process), expert human operators must revert to manual operation under strict bounds of his/her actions by a set of general operational software constraints. The Human in the Loop must execute bump-less transfer of controllers (to manual) and perform certain elementary step-by-step actions up to a known safe state. *The Human in the loop here is not only about safety, but concerns operations criticality.*

Examples of Standards concerning degrees of autonomy and decision making

The recent SAE standard J3016 (www.sae.org/autodrive) defines six levels of driving automation (0,1,2,3,4,5):

Level-0: No Automation, Level-1: Driver Assistance, Level-2: Partial Automation (The human driver performs part of the dynamic driving task and monitors the driving environment), Level-3: Conditional Automation (where the automated driving system performs the entire dynamic driving task), Level-4: driving systems monitor the driving environment and Level-5: Full Automation (where the automated driving systems monitor the driving environment).

These levels are descriptive, rather than normative and technical rather than legal.

---

[24]For a general description of this one of a kind process see R LÃsser et al. (2000): Overview of the Performance of the JET Active Gas Handling System, Fusion Engineering and Design. See also: A. Konstantellos, J.L. Hemmerich, A.C. Bell, et al.: The JET Active Gas Handling Plant Process Control System, Fusion Science and Technology, Volume 21, Number 2P2, March 1992 http://www.euro-fusionscipub.org/wp-content/uploads/2014/11/JETP98077.pdf

<u>Comment:</u> The challenge is a "bi-directional assistive interaction" i.e. the human as assistance to the automation and the automation as assurance and advisor to the human user. The simplest Human in the Loop (HitL) case has been known for at least 70 years and is the tuning of the three PID parameters on line (K_p, K_i, K_d) when the plant operators intervened in the cases of unexpected bad automatic (closed) loop performance. This is not the case in aerospace where the pilot may not attempt to retune on-line the control algorithms. The reason of this difference is mainly the long-time constants encountered in the process industries compare to aircraft dynamics and the availability of access to these parameters, (if the plant operational policy at all allows on-line interventions).

Note: For an excellent discussion about the role of cognition and control in the context of autonomy and the human in the loop, see Sections 4.6 and 4.7 of the recent IFAC report, [23] and Li, Sadigh, Sastry, Seshia [244], also Schirner et al. about HitL in CPS [208] and Finn Müller-Hansen et al. [271].

1.8.5 Sequential Process, CPS and Decision Making

1.8.5.1 General

In Cyber-Physical systems (and embedded and control systems used before), we frequently encounter decision making problems applied to sequential processes where a) the phases of some specific physical/chemical/biological phenomena are strictly sequential, because of, for example, transportation, or diffusion phenomena happening in real time and b) similar interesting phenomena also occur in computing and communications systems, for example in sequential data streaming processing (and threshold detection), recently reported by Rajeev Alur [158–161], Luis Mendes et al. [214] Soroudi and Amraee [218] as appropriate CPS decision making paradigms. Note that both cases a) and b) mentioned above, frequently use Wavelet-type transforms for DM purposes. Other related applications can be found in planning of sequential operations and in designing real time scheduling algorithms in several industrial sectors. Decision making is embedded in all phases of the execution, either as transition handover confirmation, or as abnormal exits, based on on-line and inferential measurements, see also [240] regarding sequetial applications in communications.

Figure 1.23 gives a typical flow and Table 1.5 shows examples of typical cycle times in different sequential applications of short, or very long duration.

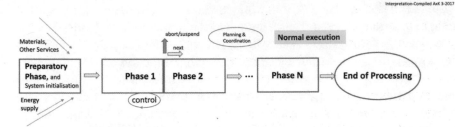

Figure 1.23   Computer-controlled sequential psysio-chemical processing (i.e. certain phases cannot be paralleled).

Table 1.5   Cycle times example

Industry	Examples: Time to build (Planning, scheduling, execution)	Typical Software size (complexity) in each product, or process	Notes
Automotive	Passenger car: 2-6 days	0.2 - 80 Mi LOCS (in product)	Embedded
Aerospace	Civil aircraft: 1-2 months	2-100 Mi LOCS (in product)	Embedded
Maritime	Cruise ship building: 1-2 years	Distributed ca. 10 Mi LOCS (process)	Production planning
Process/Plastics	From VCM to PVC production Typical cycle : 5 hours	Ca. 1 Mi LOCS (in recipe scheduling and real-time control)	Process Control (Batch) per unit
High Throughput Screening (HTS)	Robotised automation, Execution time: typical 30 Sec-3Min /plate-sample	Not available	Drug discovery

1.8.5.2 Sequential process examples

We'll present three decision making examples from the former category a) mentioned above in Section 1.8.5. The key elements to deal with are signals and the decision-making challenge is to detect, for example, when an amplitude threshold is exceeded and then trigger some automatic time-stamped messages, or actions.

• On-line Chromatography – Mass Spectrometry peak identification and peak picking decision: See [215] for an introduction. The challenge is to improve correctness of peaks evaluation by precisely locating the peaks at the detector-end in order to measure their height (which is proportional to the sample species). The execution systems can be called CPS, they are automated computer-based, executing signal processing in software (frequently here is a Wavelet transform). The physical part is a stream (gas, or liquid) with the *in situ* "sensor-detector". They deliver time stamped vectors $(a_1, a_2, a_3, \ldots, a_N)$ representing the detailed composition of the N species in the sample. The operation can be implemented in a closed loop too, but its tuning is more difficult because the measurement is very slow, therefore after a decision is reached about the composition, a delay compensation algorithm is to be executed before the feedback loop closes. The peak detection process is not periodic and not clocked, but sequential.

• Electrocardiography concerning Arrhythmia symptoms and peak pattern detection. Abas, Rodionova, Bartocci, Smolka and Radu Grosu present a smart signal analysis of a diagnosis application [146, 147]. The challenge is to identify the peaks of an Electrocardiogram (ECG) by special post processing of the signal after the standard DSP. The final result is evaluated visually by a human expert. In contrast to the previous example the signal has periodic peaks and the timing is important.

• High Throughput Screening (HTS) is a fully automated/robotised process of off-line laboratory testing of a very large number of small amounts of combined substances on plates (typical 5,000–100,000 plates per day) to be bio-chemically tested with other reacting substances. This is a multiple sequential process with several reactions/sensing/measuring stages, swift evaluation of results and multi-level decision making. HTS is recently (2015–2017) used for drugs development and pharmaceuticals accelerated automated experimentation [222–226].

Part III – Additional Topics and Concluding Remarks

1.9 Requirements Engineering and Technology Maturity Levels

Two important technological aspects are briefly discussed in this section which are also related to the emerging topics dealt with in this chapter. They are vital elements of general best-practices and engineering excellence. First, we emphasise the importance of considering activities, early in the design, both concerning the Functional Requirements (FR) and the Non-functional Requirements (N-F R) and their harmonisation and secondly to discuss the traditional Technology Readiness Level (TRL) concept which was proposed by NASA, in the early 70's vis à vis the needs of contemporary multiple technologies driven systems, like CPS, SoS and Decision Making at global and local levels. The discussion aims at raising awareness among new practitioners and researchers.

1.9.1 Requirements Engineering

Requirements span the whole spectrum from the system user/owner needs, through systems design, specific applications engineering and operations up to business and economic aspects and imperatives. The meaning of the term requirements, depends on the level/granularity of the designer/owner to express and formulate his/her initial needs for a product/process, service or improvement activity and the capability to proceed with further refinements. Requirements and requirement specifications can be prepared in plain language, structured text, up to a set of text with multi-aspect diagrams, pseudocode, preliminary software requirements and formal languages supported by appropriate tools. Intermediate expertise may also be necessary when the originators do not have the required knowledge. The originator/owner may be familiar with some initial requirements, but not with all possible system aspects and functionalities in the Cyber part (architectures, languages, programming, flexibility, boundaries etc).

FR and NFR

1. The functional requirements (FR) refer to what the system, or product is supposed to do, are usually specified by non-computer scientists/ engineers, but, at least at the beginning of the design, by domain experts (e.g. transportation, energy, manufacturing) and at some next stages by systems, control and software engineers – in the traditional meaning of the engineering disciplines. Today, since the boundaries of these disciplines are blurring, we move towards more concurrent activities to achieve coherent and correct designs [288].

2. The Non-functional requirements (NFR) for a system, or product are also called extra-functional requirements. The term NFR mainly refers to software and hardware aspects, such as interoperability, dependability and safety, scalability, security, robustness, timeliness, performance and upgradability and means *how the system will do what the functional requirements specify* [289].

There have been various opinions and practices concerning the separation, or the concurrency in addressing FR and NFR. Regarding embedded systems design for example, J. Sifakis stated that "Separation of Concerns – Separately addressing functional from extra-functional requirements is essential from a methodological point of view"[25]. See also [227–230].

In any case overarching constraints in all these considerations are time, resources and cost. Good requirements from the start and clear specifications are essential for successful next steps in design and implementation of systems.

---

[25]Joseph Sifakis: A Framework for Component-based Construction, 2005, http://www-verimag.imag.fr/sifakis/RECH/SEFM05-sifakis.pdf

Elementary Example: The functional requirements for a basic control system are (in simplified form):

$FR = \{$read a temperature measurement from a sensor, apply a filter for noise rejection, compare this value (ϑ) with the stored set point – i.e. the desirable temperature (ϑ_s) – then, use the difference $(\vartheta - \vartheta_s)$ to execute a control algorithm and finally send the result of this calculation to the system output channel in order adjust an actuator, usually a heater$\}$.

$NFR = \{$how fast is the end-to-end computation from sensor to actuator done in the worst case, what is the accuracy of the calculations, how secure is the connection/link from the sensor to the computing system, what will happen and what the system should do if the sensor, or the whole control system fails for a short period, how is the state of the heater monitored, etc$\}$.

The functional part is of prime importance for the system design and execution and requires domain-knowledge, and furthermore matching of the physical (dominant FR) and the cyber parts (dominant FR and NFR).

Concerning large scale systems (including SoS which are frequently heterogeneous), the FR-NFR topic is more demanding because: it requires systematic requirement capturing tools, management and decision making support, alignment of technology maturity levels, as well as software tools for coordination purposes and compliance with standards. In this context "Contract-based" design methods are the ultimate approach for ensuring compliance and traceability at later stages [121, 122].

Furthermore beyond Systems and Software Engineering considerations, in case of creation of cyber and physical parts, in a CPS context, in particular for grass root designs (products, or contracted system designs), or whenever design is not yet fixed, a cooperation, or concurrent consideration of FR and NFR of the human-made parts and of the being-created-CPS may be advantageous. There are also some implications of the correct FR-NFR interaction during initial design, on the technology maturity levels discussed in Section 1.9.2.

In summary: Where possible a *cooperative FR and N-FR analysis and cross-profiling*, under a coordinating supervision may be advantageous, beyond the classic concurrent engineering approach usually targeting simultaneous process and control systems design of the past. This joint FR-NFR effort should be applied both at systems and application engineering levels. Furthermore, the partition of the implementation of systems functions and applications into hardware, software and combined cases needs special attention regarding FR and NFR because it needs complementary expertise. Criticism about the non-clarity of some NFR definition is provided in [290].

1.9.2 Is TRL Sufficient for CPS and SoS/E?

Technology maturity assessment is an avenue that engineers and program managers utilize to make critical decisions about the probability that a technology can contribute to the success of a system [236].

Assessing the maturity level of technologies level and cross-disciplinary fertilisation

The use of Technology Readiness Level (TRL) and the recently proposed complementary scales, such as Integration Reediness Level (IRL), Manufacturing Readiness Level (MRL), System Readiness Level (SRL), and with the same abbreviation Software Readiness Level (SRL), is appropriate for comparing alternatives, assessing suitability of interfacing and linking subsystems (constituent systems in SoSE) and evaluating economic viability of offers. If we call them TRL+ they are therefore promising Decision Making Support tools, which can, for example define progress/delays in large scale engineering undertakings.

The technology readiness level (TRL) scale was introduced by NASA in the 1970s as a tool for assessing the maturity of technologies during complex system development [26].

In 1995, NASA published a refined 9-point scale, along with the first detailed descriptions of each level. The TRL scale is designed to assess each component technology independently; however in reality, the components are integrated to work as a complete system[27].

Example: The reports about significant delays in the software testing, for the completion of the F-35 fighter aircraft [287], refer to gaps and inabilities of today's systems tools to ensure time compliance. Newly developed complex CPS-type processes from software methods to commercial grade tools seem not immediately ready for integration [231–237].

Comment: One of the sources of execution delays, is the mismatch between requirements and the developed (for use) solutions. This in its turn, justifies the R&D efforts to create rigorous methods and support tools for requirements capture. See for example a very detailed report [121] and [122] – part of

---

[26]https://www.nasa.gov/directorates/heo/scan/engineering/technology/txt_accordion1.html

[27]Kyle Y. Yang, James Bilbro (MIT Lincoln Laboratory): Technology Integration and Improved Technology Maturity Assessments.

the result through several European Projects (SPEEDS, DANSE, and earlier ASSERT by ESA). A criticism of such new approaches from a practitioner's point of view refers to the ability of industrial companies and development teams to understand and use the new rigorous methods. Certainly, the next generation "systems education" should help create the right engineers for the future.

Important note: To move a technology, for example from a TRL 3 to a TRL 8 may take several years in practice.

TRL approach drawbacks

TRL provides appropriate technology readiness assessment mainly of components developments. For this reason, TRL has been criticized concerning its applicability with respect to large complex engineering developments. Empirical evidence also showed that programmes with TRL 5 and below, exhibit significant time delays and cost overruns when used. The ideal starting level is TRL 7 for industrial success, while several stakeholders would like to have TRL 9. This is confirmed in a European Survey by Leitão, Karnouskos et al. (2016) [33].

What is not adequately addressed in TRL, are aspects like integration, risks and system level readiness, heterogenous platforms and capabilities for insertion of additional technologies at later times, because TRL was not created to handle such issues [234]. Now, since CPS and SoSE developments are based, in their majority, on multiple and heterogeneous technologies with emphasis on total system design and integration, the previous comment on the weaknesses of TRL is applicable specifically to these systems.

Proposed improvements by Industry

To date, improvements have been proposed by industry. The most prominent ones are Integration Readiness level (IRL) [233], System Readiness Level (SRL) [234] and Manufacturing Readiness Level (MRL) [235]. The first two are very relevant for large scale CPS and SoS/E designs and projects.

Integration Readiness Level (IRL): In a complex system development, we are dealing with several TRLs. Integration is understood here as the combination-interfacing and coordination of separate components. This first new term, IRL includes seven levels. The lowest is No 1 covering the maturity of interfacing/

physical connection to allow sufficient characterisation. Interesting is level 3 which requires compatibility via a common language or lexicon of terms. Level 7 refers to the verification and validation of the integrated technologies so that the system is actionable. The typical example mentioned is when two technologies with IRL=8 and IRL=9 respectively were combined in a Mars Climate Orbiter of NASA. It required an IRL of minimum 5 to cover integration vulnerabilities in software.

System Readiness Level (SRL): This development metric addresses the phases of designed framework for Life Cycle Management. SRL includes five five levels: Level 1 deals with refinement of initial concept through a strategy, Level 2 considers the reduction of technological risk to build a full system, Level 3 system development demonstration to ensure interoperability, safety and security, Level 4 achieves operational capability to meet mission/project needs and Level 5 addresses sustainability of the system in its total life cycle.

Enhancements: Sauser et al. proposed an empirical function linking the three terms (*System, Integration* and *Technology*) SRL = $f(TRL, IRL)$ [232, 233] with average aggregated $SRL = \frac{1}{N} \times [(\frac{SRL_1}{m_1}) + (\frac{SRL_2}{m_2}) + \ldots + (\frac{SR_k}{m_k})]$, where N is the total number of technologies and m_k reflects the number of integrations per sub-system, or constituent system (in SoS jargon) with own SRL_k. Next, Sean Ross [237] enhanced the Sauser-metric mentioned above, to include Manufacturing Readiness Level (MRL) and proposed $SRL_g = g(TRL, IRL, MRL)$. He stated a valuable comment that frequently "The competing pitfall in system development is the premature advancement of a technology to the next level of development in advance of its interfaces". Interfaces means integration, therefore "*We can do a better job by minimizing the gap between interface, manufacturing, and technology maturity*". This statement is related to the delays in the completion of major U.S. defence projects[28].

Comment: The careful management of Maturity Levels (*RL) towards integration, is important and in line with the key CPS objective for tight integration (between Cy-Phy parts), but also points to the critical situations arising from early adoption of fresh technologies (e.g. Pre-procurement policies in the US and Europe).

---

[28] http://www.gao.gov/assets/670/661842.pdf

1.10 The System of the Future (SoF) – Food for Thought

1.10.1 The Dreams and Visions of the Systems R&D Communities

Since CPS addresses in detail the design process and ensures certain properties when the systems are built, much more than the other new topics, CPS probably meets the dreams of several distinguished scholars and industry leaders in systems, control design and their manufacturing who have expressed such views within the underpinning domains before and towards CPS. **Their visions have been e.g. about a unified systems theory, systems with rich functionalities, highly dependable and resilient systems, fault tolerant controls and extremely powerful microelectronics to become low cost multi/many core, certified for mixed critically and designed with rigorous systems methods for robust real time operations.** For example, the contributors in cross-domain brainstorming meetings who providedideas and suggestions on various occasions beyond projects, to the genesis of the European and International Networked Embedded and Control systems area (NE&C), the subsequent private-public partnership ARTEMIS/ECSEL, and the S&T activities in the new CPS and SoS/E domains. Among them are: Frank Algöver, Panos Antsaklis, Karl-Eric Årzén, Karl Johan Åström, Thomas Bak, Eric Bantegnie, John Baras, Alberto Bemporad, Koen De Bosschere, Manfred Broy, Alan Burns, Eduardo Camacho, Elias Carayannis, Armando Colombo, Patrick Cousot, Werner Damm, Jerker Desling, Steven Ding, Zeljko Djurovic, Maria di Benedetto, Zoe Doulgeri, Sebastian Engell, Rolf Ernst, John Fitzgerald, Laila Gide, Peter Groumpos, Michael Henshaw, Sandra Hirche, Paul van den Hof, Alf Isaksson, Nick Karkanias, Stamatis Karnouskos, Michel Kinnaert, Hermann Kopetz, Stefan Kowalewski, Francoise Lamnabhi-Lagarrigue, Kim Larsen, John Lygeros, Hugo de Man, Lorenzo Marconi, Manfred Morari, Richard Murray, Henk Nijmeijer, George Pappas, Thomas Parisini, Ron Patton, Maria Prandini, Anders Rantzer, Alberto Sangiovanni – Vincentelli, Tariq Samad, Ricardo Sanz, Shankar Sastry, Bart de Schütter, Jan Van Schuppen, Joseph Sifakis, Janos Sztipanovits, Haydn Thomson, Martin Törngren, Mateo Valero, Tulio Vardanega, Zoran Vukic, Carlos Canudas de Wit and several others included in the related meetings reports.

Have all these dreams been realised today? Probably not yet, but it seems the related communities are on the right direction because, at least the needs and the gaps are now better understood.

1.10.2 The System of the Future (SoF)

Figure 1.24 illustrates an impression of what a large-scale System of the Future (SoF) might look like. Macroscopically the system of the future may resemble a SoS and microscopically a well-controlled CPS, while the glue may be the future Internet of everything. This "super-SoS" may consist of several "co-existing" systems, living and operating between two worlds (real and virtual):

- a structured group of predictable, superimposed environments, creating sequential, parallel, or mixed events which would affect some, or all systems within a period;
- a fuzzy horizon full of unknown and unpredictable uncertainties, i.e. difficult to a-priori model, but influencing the states of the intercommunicating systems.

Between these two worlds, systems of diverse size, nature, human and social content will be operating and living semi-autonomously, or autonomously under light but fair supervision, which will be setting strategic goals, as well as safety, security limits and ensuring consistent global time by orchestrating the execution of possibly conflict-free harmony, using cognitive agent-based communications and interactions. Each of the systems S_i could be characterised by an identity metric Q_i:

$S_i \rightarrow Q_i : Q_i|t =$ (*goals, availability, capabilities, reactivity, dynamic behaviour, past history, status, global time*).

For example, status may be one of the following: unknown, stand-by, in conflict (i vs. j), stand-alone active normal, cooperating (j, k, n), candidate,

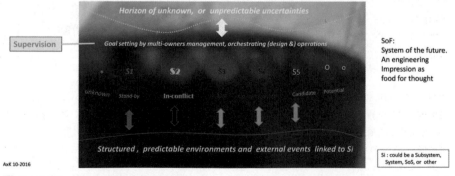

Figure 1.24   The System of the Future (illustration from a macroscopic organisational point of view).

candidate under validation. *Utopia, or Possibility?* The future will show how many of the new and forthcoming ideas and technologies will be realised and successfully taken up. But certainly, nobody expects to design CPS, IoT and SoS to last for 300 years, or build systems which will never fail and be flexible enough to meet all known and the emerging requirements in the future.

1.10.3 Further Examples of Systems Topics for Future R&D Activities

Beyond the CPS oriented challenges refered to in Section 1.5.11 of this chapter, summarised below are some additional systems ideas emerging for example from the European Union, such as the work-program 2016–17 [249], the International Federation of Automatic Control (IFAC) strategic report on systems and control (2017) [23], the OECD vision 2016 [128] the NSF CPS solicitations [1] and numerous other papers and reports on the new topics, from academia, R&D projects, industry and individual researchers [1–4, 34, 105].

- **Universal Frameworks (for CPS and SoS)** such as BIP (Behaviours, Interactions, Priorities) & Contract-based Design (proposed by several EU and CH groups) "to become a broad platform and design language, beyond *ML, with rich tool sets for studying, modelling and designing industrial systems, – these needs to overcome mixed criticality certification obstacles and certain scalability shortcomings". (Originators: J. Sifakis, S. Bensalem, A. Benveniste, Bozga, Alberto Sangiovanni-Vincentelli, Janos Stzipanovits).
- **Integrated Decision Support Methodologies** (proposed by R&D teams in Decision Making and Control) "to extend basic probabilistic and geometric (e.g. graphs) methods to cope with many highly-distributed agents towards computationally scalable solutions for e.g. Decentralised Markov Decision Processes modelling and multi-level optimisation problems. Applications to include multiple aircrafts and UAV presence – collision avoidance challenges". (Compiled by the author. Sources: on going work at MIT, CMU, Stanford, FAA and SESAR R&D activities in Europe).
- **Strengthen CPS Foundations**: "encourage new scientific concepts to model and understand the interdependencies between human-made and natural world (including rivers, lakes, forests, seas, oceans), while the Cy-Phy components need new architectural models redefining structure and functions for greater assurances of safety, security, scalability

resilience of the integrated CPS and to safeguard open interfaces under modularity, trouble free interoperability and rapid verification of components and of the synthesized end-to end systems" (based on NIST [20], NSF [1, 26, 43, 104, 147, 213]).

- **Elaborate Contemporary Complexity and Technology Maturity Metrics**: "Support cross disciplinary definition of appropriate CPS and SoS complexity and new technology maturity metrics and promote their take up based on a balanced use of mutual state of the art results of comparable TRL (proposed by the author based on [234, 235]).

Encourage CPS and SoSE R&D to support 5G Global networks and beyond.

1.11 Concluding Remarks

It is appropriate to mention, that the upper levels of large systems requiring DM/DeP could be addressed also under a comprehensive Systems Engineering and a Systems of Systems perspective in conjunction with CPS thinking for the individual systems design. This would be useful when upgrading large scale engineering projects. Furthermore, the role of the Physical/Human-made "non-ICT" systems should be emphasised, because these too must be conceived, designed and implemented, and where possible, concurrently with the "Cyber" parts, or as joint creations.

The major documents defining CPS, SoS and IoT, frequently refer to applications at very high level and therefore create the impression that they all do the same thing: for example, "IoT for energy and transportation", "CPS for energy and transportation", "SoS/E for energy and transportation". This should not be interpreted as overlap. In fact, while there has been a substantial amount of initial hype and despite the opposing claims in the debates, the practical applications of the new technologies make matters simpler than the cross-domain discussions. Industrial applications for example, will have less problems to select between IoT, SoSE and CPS-based solutions if they need to.

As indicated earlier: IoT is network-centric and more applications oriented, while CPS is cross cutting system-design centric. CPS addresses several fundamental and specific systems challenges across all sectors, which are not involved in, or bother IoT at the same level of granularity and depth. Primarily IoT addresses several other fundamental technological issues (e.g. air interfaces, antennas, self-configuration, light protocols, battery operated nodes) which are not core activities in CPS, but CPS could contribute to these developments from a design methodology perspective. Several CPS rigorous design and verification methods can benefit IoT developments. CPS may use IoT as one of its system platforms, or probably even as a major CPS substrate for example using Industrial IoT. Therefore, CPS and IoT approaches are not only complementary, but can both cooperate towards synergistic solutions.

A system of systems (SoS) is a large-scale system, a large CPS is also a large-scale system. However, a SoS by definition and by convention, exhibits specific behaviours probably not applied to all large CPS and on the other side, SoS Engineering deals more with the interactions, organisation, coordination and management of the SoS constituent systems, than about their individual inherent designs. CPS (including DM) could provide SoSE with systematic methods for estimating, evaluating, simulating and optimising behaviour, resources and expenditure. Smaller SoS may be CPS and large CPS may exhibit SoS phenomena. Some SoS concepts could help CPS regarding for example emergence and high level stabilisation, although CPS designs cannot be based only on emergent properties.

Figure 1.25 schematically depicts the new domains and the key synergistic bridges between them. To realize and benefit from these, stronger cooperation among the related S&T fields is necessary, since they usually target similar technological, socio-economic and sectoral/business objectives.

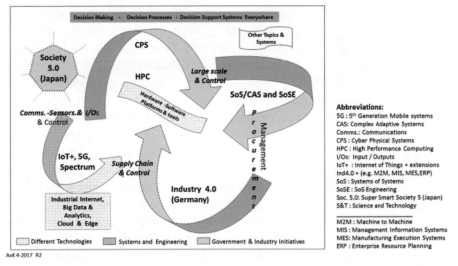

Figure 1.25 The emerging high level S&T systems concepts, technologies and initiatives.

Finally, while questions like *"what is CPS and what not"* and *"what is the boundary between Cyber and Physical parts"* are reasonable, they are not critical. A more useful point to pose would be for example: **"For a given system, or concept X, what CPS methods, approaches and tools would be suitable for its design, analysis, engineering and production, or for possible improvement purposes for existing systems?"**

The topics discussed in this chapter cover a broad set of system concepts and techniques, as well as improvements of earlier results and methods. All aim to some degree at "better" systems designs and operations. In many practical applications not only one but (unfortunately) several of them are necessary to be addressed ideally by multidisciplinary S&T experts and businesspeople. Furthermore, each of the topics is touched on its surface only. There is, for each of them, a plethora of sub-topics and underpinning methods continuously evolving and technologies not mentioned at all. Some of these may however become the dominant R&D themes and approaches of the future.

1.12 Summary

This introductory chapter included three parts: a) systems, b) decision-making and c) requirements engineering. In the first part, we endeavoured to define systems and their interactions and to focus on the emerging topics CPS, IoT, SoS/E, 5G and the policy initiatives Industry 4.0, Society 5.0 and similar ones in Korea, US and Europe. The purpose was to outline the key issues, mainly from an engineering R&D and practical point of view, and pose constructive and some rhetoric questions as food for thought for further actions. It summarises, in compact form, key S&T results and methods for the newcomers to system design, implementation or operation. The presentation is by no means exhaustive, but broad and cross-cutting.

On one side the problems encountered are many; they are complex, and frequently difficult to correctly formulate, interpret and tackle. On the other side, also the existing physical systems and human-made systems are themselves, many, different, complex and finally also the used and new/proposed solutions, methods and tools – however smart and innovative they may be – they need further refinements and improvements through collective and individual R&D efforts. The successful developments and ultimate efficient and effective take up of the results, depend on the degree of interdisciplinary and cooperative work of teams and the support on behalf of all stakeholders involved.

To facilitate the profiling and categorisation of systems, either single ones, or multiple systems in an environment, two intuitive metrics are discussed one based on scale, behaviour and structure and the second reflecting the nature of a system based on its content proximity to four spaces namely a combination of {*cyber, human, social* and *physical*}. This is applied to CPS as basic classification tool.

Even though there are no universally accepted definitions of CPS, IoT and SoS, they all have distinct S&T assets and a strong role to play in the future. It was proposed in this chapter to use any simple definition for each of them, accompanied by a small set of key requirements which will aim at highlighting the advantages over earlier approaches, designs and solutions. The new domains appear useful not only to big sectors such as energy, transportation, robotics and health care, but to smaller scale attractive applications such as in forestry, maritime & ocean engineering, shipbuilding, arts, sports and music synthesis.

The second topic of this chapter is Decision making (DM). Its scientific methods facilitate reaching quantitative converging solutions to problems of selection among many options, in processes under variable deterministic, or under non-deterministic/stochastic uncertainties e.g. in CPS, IoT and SoS. Decision Making (DM) difficulties may increase with the scale of the systems. Several examples are presented as illustrations of DM from threshold detection in Gas Chromatography, EEG, EEG and ECG with the help of special Integral Transformatios, multiple cranes movements in congested construction sites (as CPS) and the forthcoming ACAS-X mid-air collision detection system as DM, or CPS. Several references are included on the origins and state of the art.

In the third topic, we have discussed the role of requirement engineering as another cross-cutting topic very important to ensure successful systems in any of the other domains. This is a dual topic addressing both functional requirements and non-functional requirements relevant for example to physical, cyber parts and the combined CPS as well as SoS engineering. Under this topic, we also commented that the TRL maturity indicator introduced more than 30 years ago, does not adequately cover CPS and SoS. This is consistent with the proposed in the literature System and Integration Maturity levels (SRL/IRL) which are more suitable for example for multiple technologies and heterogeneous systems. These indicators are responsible not only to determine the maturity level of technologies but to assist in the combination of technologies and the prediction of possible mismatches. It is up to the relevant communities to advance the issue of improving TRL.

In brief, IoT is a set of networking-friendly technologies and a market driven area, SoSE is an engineering, organisation-friendly domain, while SoS (also CAS, ULSS) are large scale system classes, CPS is a broad integrated systems and control design conceptual umbrella with the potential to become the systems domain of the future, Ind 4.0 is a manufacturing oriented initiative, and 5G is the next generation wireless technology. Progress in these new fields is expected, first through the continuation of R&D and innovation activities in their already established S&T areas and in the underpinning sub-domains, and secondly through totally new ideas and methods, probably coming out of novel mathematical theories such as co-algebras, special lattices, random trees, functional transformations and new concepts. A third route could be the cross-fertilisation of methods and solutions from other areas of science and technology.

The fascinating systems world is full of opportunities, difficulties and surprises. Perhaps, the three forms of existence defined by Roger Penrose [242] as "The physical, the mental and the Platonic mathematical – as entities belonging to three separate worlds – and the profound mysteries in the connections between them, ... together with Beauty, Morality and Truth" are meeting the emerging realities of data, information and human ambitions. Even the hot topics of today may be replaced by evolving new ones like the expected transition from energy efficiency to data efficiency. Within the surrounding us multiple complexities, CPS, IoT and SoS/E are strong enablers of further innovation.

This chapter touched only the surface of some of the many interesting new topics, giving introductory and simple paradigms from science and industry to encourage further R&D explorations, engagements and more industrial activities in any useful interaction among the Cyber, Social, Human and Physical worlds.

Acknowledgements

The author would like to thank the editors Prof. Dimitri Soudris (NTUA, Athens), Prof. Kostas Siozios (University of Thessaloniki) and Prof. Elias Kosmatopoulos (Democritus University of Thrace) for encouraging me to write this free-style chapter and for their valuable advice. I'm grateful to Junko Nakajima and colleagues at River Publishers for their help in organising the text and the references.

I would also like to thank Prof. Michael Henshaw, University of Loughborough UK, for his comments on the deep aspects of Systems of Systems and for bringing to my attention, as leading member of the scientific team, the report of the European Parliament on the Ethical issues of Cyber-Physical Systems.

My gratitude to Prof. Hermann Kopetz, Technical University of Vienna is boundless for sharing with me his ideas on robust industrial systems and networks and for providing information about current applications of the TTA-driven methods, bridging gaps between theory and engineering.

References

[1] NSF CISE Cyber-Physical Systems (CPS). (2016–2017). *Synopsis, Solicitation 17-529*. Availbale at: https://www.nsf.gov/funding/pgm_summ.jsp?pims_id=503286

[2] National Science and Technology Council (USA). (2016). *"The National Artificial Intelligence Research and Development Strategic Plan,"Networking and Information Technology Research and Development Subcommittee*. Available at: https://www.nitrd.gov/PUBS/national_ai_rd_strategic_plan.pdf

[3] European Commission (2016). *Digitizing Europe Initiative*. Available at: https://ec.europa.eu/digital-agenda/en/digitising-european-industry

[4] European Commission. (2017). *Cyber-Physical Systems*. Available at: http://ec.europa.eu/digital-agenda/en/cyber-physical-systems

[5] Japan Report (2015). *"The 5th Science and Technology Basic Plan" – "Super Smart Society" (Society 5.0), Council for Science, Technology and Innovation Cabinet Office, Government of Japan (December 18, 2015)*. Available at: http://www8.cao.go.jp/cstp/kihonkeikaku/5basic plan_en.pdf

[6] Harayama, Y. (2016). *600 Trillion Yen GDP Target STI Policies for Moving Toward Society 5.0!*. Available at: http://fpcj.jp/wp/wp-content/uploads/2016/07/f2d3eec7bf7678840f8adf2ca8000b05.pdf

[7] Vision 2025. (2000). *R&D Plan, Korea, Long Term Plan for Science a Technology Development*. Available at: http://unpan1.un.org/intradoc/groups/public/documents/APCITY/UNPAN008040.pdf

[8] Vision for Science 2025. (2014). *Vision for Science, Choices for the Future, The Hague*. Hague: Ministry of Education, Culture and Science of the Government of the Netherlands.

[9] China-25 initiative. *Possibilities and Challenges for Industry 4 in China*.

[10] Lawrence, M. (2016). *Report Future Proof Britain in the 2020s, IPPR 2016 Institute for Public Policy Research*. Available at: http://www.ippr.org/files/publications/pdf/future-proof_Dec2016.pdf? noredirect=1

[11] Lee, E. A., and Seshia, S. A. (2015). *Introduction to Embedded Systems, A Cyber-Physical Systems Approach*, 2nd Edn. University of California at Berkely. Available at: https://LeeSeshis.org, EE&CS

[12] Alur, R. (2015). *Principles of Cyber-Physical Systems*. Cambridge, MA: The MIT Press.

[13] Merwedel, P. (2011). *Embedded Systems Design, Embedded Systems Foundations of Cyber-Physical Systems*, 2nd Edn. Berlin: Springer.

[14] Kochenderfer, M. (2015). *Decision Making Under Uncertainty*. Cambridge, MA: MIT Press.

[15] Kopetz, H. (2011). *Real-Time Systems: Design Principles for Distributed Embedded Applications*. Berlin: Springer Verlag.

[16] Bondavalli, A., Bouchenak, S., and Kopetz, H. (eds) (2016). *Cyber-Physical Systems of Systems Foundations "A Conceptual Model and Some Derivations: The AMADEOS Legacy 2016"*. Berlin: Springer.

[17] Jamshidi, M. (2009). *Systems of Systems Engineering: Principles and Applications*. Boca Raton, FL: CRC Press.

[18] Lee, E. A., and Varaiya, P. (2011). *Structure and Interpretation of Signals and Systems*, 2nd Edn. Boca Raton, MA: Addison-Wesley.

[19] Fradkov, A. (2006). *Cybernetical Physics – From Control of Chaos to Quantum Control (Understanding Complex Systems)*. Berlin: Springer. Available at: www.ipme.ru/ipme/labs/ccs/

[20] NIST-Cyber Physical Systems Public Working Group. (2016). *"Framework for Cyber-Physical Systems."* Available at: https://s3. amazonaws.com/nist-sgcps/cpspwg/files/pwgglobal/CPS_PWG_Fram ework_for_Cyber_Physical_Systems_Release_1_0Final.pdf

[21] Kim, K.-D., and Kumar, P. R. (2014). *An Overview and Some Challenges in Cyber-Physical Systems*. Available at: http://cesg.tamu.edu/ wp-content/uploads/2014/09/An-Overview-and-Some-Challenges-in-Cyber-Physical-Systems.pdf

[22] Babiceanu, R. F., and Seker, R. (2016). Big data and virtualization for manufacturing cyber-physical systems: a survey of the current status and future outlook, *Comput. Ind.* 81, 128–137.

[23] Lamnabhi-Lagarrigue, F., Annaswamy, A., Engell, S., Isaksson, A., Khargonekar, P., Murray, R., et al. (2017). Systems & control for the

future of humanity, research agenda: current and future roles, impact and grand challenges. *Ann. Rev. Control* 43, 1–64.

[24] Energetics Incorporated Columbia, Maryland for the National Institute of Standards and Technology. (2013). *Foundations for Innovation in Cyber-Physical Systems, Workshop Report*, Available at: https://www.nist.gov/sites/default/files/documents/el/CPS-WorkshopReport-1-30-13-Final.pdf

[25] Shi, J., Wa, J., Yan, H., and Suo, H. (2011). *A Survey of Cyber-Physical Systems*. Available at: https://www.researchgate.net/publication/22893 4884_A_Survey_of_Cyber_Physical_Systems

[26] Khaitan, S. K., and McCalley, J. D. (2014). Design techniques and applications of cyber physical systems: a survey. *IEEE Syst. J.* 9, 350–365.

[27] Abbaspour, A. S., Inam, R., and Hansson, H. (2015). "A survey on testing for cyber physical system," in *Testing Software and Systems. Lecture Notes in Computer Science*, Vol. 9447, eds K. El-Fakih, G. Barlas, and N. Yevtushenko (Berlin: Springer), 194–207.

[28] Abulamddi, M. (2017) A survey of approaches reconciling between safety and security requirements engineering for cyber-physical systems. *J. Comput. Commun.* 5, 94–100.

[29] Ye, H. (2015). Security protection technology of cyber-physical systems. *Int. J. Sec. Appl.* 9, 159–168.

[30] Larrucea, X., Combelles, A., Favaro, J., and Taneja, K. (2017). *Software Engineering for the Internet of Things*. Available at: https://www.computer.org/csdl/mags/so/2017/01/mso2017010024.pdf

[31] Road2CPS (EC project workshop). (2016). *Smart Cyber-Physical Systems Clustering and Communication Event*. Vienna: Agenda.

[32] Heemels, M. (2017). *Cyber-Physical systems and Control Systems Research*. Available at: https://heemels.tue.nl/research/cyber-physical-systems

[33] Leitão, P., Karnouskos, S., Ribeiro, L., and Lee, J. (2016). Thomas strasser and armando w. colombo, smart agents in industrial cyber-physical systems. *Proc. IEEE* 5, 1086–1101.

[34] Davies, R. (2015). *The Internet Of Things Opportunities And Challenges, Eprs – European Parliamentary Research Service*. Available at: http://www.europarl.europa.eu/RegData/etudes/BRIE/2015/55701 2/EPRS_BRI(2015)557012_EN.pdf

[35] Vermesan, O., and Friess, P. (2016). *Internet of Things – From Research and Innovation to Market Deployment*. Aalborg: River Publishers.

[36] CyPhERS. (2014). *Cyber-Physical European Roadmap & Strategy, State of the Art*. Available at: http://www.cyphers.eu/sites/default/files/ D2.1.pdf

[37] Tzanev, A. (2013). *Modeling and Simulation of Systems of Systems "a Survey, University of Chemical Technology and Metallurgy (UCTM)" Sofia, Dept. of Industrial Automation, Report, 2013*. Available at: http://www.cit.iit.bas.bg/cit_2013/v13-2/10341-volume13_issue_2-01_paper.pdf

[38] Desling, J. (2015). *Building Automation Systems from Internet of Things, Luleå University of Technology, Report presentation*. Available at: http://www.arrowhead.eu an Artemis project and http://www.ProcessIT.eu

[39] Barot, V., Henshaw, M., Siemieniuch, C., Sinclair, M., Lim, S. L., Henson, S., et al. (2013). *State of the Art SoS, Report of the EU-US project T-AREA-SoS, Loughborough University*. Available at: https://www.tareasos.eu/docs/pb/SoA_V3.pdf

[40] AMADEOS EC Project. (2014). *Commonalities and Requirements, – Architecture for Multi-criticality Agile Dependable Evolutionary Open System-of-Systems*. Available at: http://amadeos-project.eu/wp-content/uploads/2014/10/AMADEOS_WP1_D1.1_v3.11final-version. pdf

[41] Thramboulidis, K. (2015). A cyber-physical system-based approach for industrial automation systems. *Comput. Ind.* 72, 92–102.

[42] Lee, E. A. (2007). *Cyber Physical Systems: Design Challenges, ISORC 2008*, Available at: http://www.cs.virginia.edu/son/cs851/ papers/CPS.challenges_ISORC08.pdf

[43] Smart Cyber-Physical Systems " Concertation Event (2017). *Organised by three EC Funded Projects Road2CPS, TAMS4CPS and sCorPiuS, (Note: Projects Presentations)*. Available at: http://road2cps.eu/ events/?p=1063

[44] Geisberger, E., and Broy, M. (eds) (2014). *Living in a Networked World. Integrated Research Agenda CyberPhysical Systems (agendaCPS) (Acatech Study)*, Munich: Herbert Utz Verlag 2014. Available at: http://www.cyphers.eu/sites/default/files/acatech_STUDIE_agenda CPS_eng_ANSICHT.pdf

[45] Ecsel Joint Undertaking. (2016). *ECSEL Multi-Annual Strategic Plan 2016 (EU), OJ L 169/152, 7.06.2014.* Available at: http://www.ecsel.eu/web/downloads/documents/ECSEL-GB-2014-06-MASP_v2.pdf

[46] Pappas, G. J., Tomlin, C., and Sastry, S. (1996). "Conflict resolution for multi-agent hybrid systems," in *Proceedings of the 35th Conference on Decision and Control Kobe, Japan December 1996.* Available at: http://robotics.eecs.berkeley.edu/sastry/pubs/Pdfs.1998%26Before/PappasConflicResolution1996.pdf

[47] Yang, L., and Kuchar, J. (1997). *Prototype Conflict Alerting Logic for Free Flight, AIAA 97-0220, January 6–10, 1997, Reno.* Available at: https://ntrs.nasa.gov/archive/nasa/casi.ntrs.nasa.gov/19980019279.pdf

[48] Tomlin, C., and Lygeros, J. (2000). A game theoretic approach to controller design for hybrid systems. *Proc. IEEE* 88, 949–970.

[49] Colombo, A., and Del Vecchio, D. (2005). *Efficient Algorithms for Collision Avoidance at Intersections.* Available at: https://scripts.mit.edu/ ddv/publications/PaperHSCC5.pdf

[50] Schebesta, H., Contissa, G., Sartor, G., and Masutti, A. (2015). *Design According to Liabilities: ACAS X and the Treatment of ADS-B Position Data, Fifth SESAR Innovation Days, 1st 3rd December 2015.* Available at: http://www.sesarinnovationdays.eu/files/2015/Papers/SIDs_2015_paper_39.pdf

[51] Kochenderfer, M., Holand, J., and Chryssanthakopoulos, J. (2012). Next-generation airborne collision avoidance system. *Lincoln Lab. J.* 19:2012.

[52] Holand, J., Kochenderfer, M., and Olson, W. (2013). Optimizing the next generation collision avoidance system for safe, suitable, and acceptable operational performance, 10th USA-Europe ATM R&D Seminar. *Air Traf. Control Q.* 21

[53] Mueller, E., and Kochenderfer, M. (2016). "Multi-rotor aircraft collision avoidance using partially observable markov decision processes," in *Proceedings of the AIAA Modeling and Simulation Technologies Conference, AIAA Aviation, June 2016,* Reston, VA.

[54] Ghorbal, K., Jeannin, J-B., Zawadzki, E., Platzer, A., Gordon, G. J., and Capell, P. (2015). Airborne collision avoidance system as a cyber-physical system. *INCAS Bull.* 7, 129–144.

[55] Jeannin, J-B., Ghorbal, K., Kouskoulas, Y., Schmidt, A., Gardner, R., Mitsch, S., et al. (2016). A formally verified hybrid system for safe

advisories in the next-generation airborn collisionn avoidance system. *Int. J. Softw. Tools Technol. Trans.* 1–35.

[56] Chen, M., Shih, C.-Y., and Tomlin, C. (2016). Multi-vehicle collision avoidance via hamilton-jacobi reachability and mixed integer programming. *IEEE Conf. Decisi. Control.*

[57] NETALERT – The Safety Nets newsletter. (2013). *ACAS-X the Future of Airborne Collision Avoidance, Eurocontrol, Netalert.* Available at: http://www.skybrary.aero/bookshelf/books/2390.pdf

[58] Chomik, G. (2014). The future of collision avoidance "ACAS X,". *Int. J. Eng. Trends Technol.* 39, 284–287.

[59] Netjasov, F., Vidosavljevic, A., Tosic, V., Everdij, M. H., and Blom, H. A. (2013). Development, validation and application of stochastically and dynamically coloured Petri net model of ACAS operations for safety assessment purposes. *Transp. Res. C Emerg. Technol.* 33, 167–195.

[60] Netjasov, F., Vidosavljevic, A., Tosic, V., Everdij, M., and Blom, H. (2013). *Stochastically and Dynamically Coloured Petri Net Model of ACAS Operations.* Available at: http://www.icrat.org/icrat/seminarCon tent/pdf/Stochastically.pdf

[61] Tanga, J., Piera, M. A., and Guasch, T. (2016). *Coloured Petri Net-Based Traffic Collision Avoidance System Encounter Model For The Analysis of Potential Induced Collisions.* Available at: https://upcommons.upc.edu/bitstream/handle/2117/85742/TGUASCH%20re vised%20version.pdf

[62] Air, A. P. (2015). *Revising the Airspace Model for the Safe Integration of Small Unmanned Aircraft Systems.* https://images-na.ssl-images-amazon.com/images/G/01/112715/download/Amazon_Revising_the_Airspace_Model_for_the_Safe_Integration_of_sUAS.pdf

[63] Welch, A. (2015). *A cost-benefit analysis of Amazon Prime Air, 2015.* Available at: http://scholar.utc.edu/cgi/viewcontent.cgi?article=1051& context=honors-theses

[64] Tuan, A. Y., and Shang, G. Q. (2014). Vibration control in a 101-storey building using a tuned mass damper. *J. Appl. Sci. Eng.* 17, 141–156.

[65] Books Llc. (2010). *Gotthard Base Tunnel.* Available at: https://en.wiki pedia.org/wiki/Gotthard_Base_Tunnel2016

[66] Dyson Blade-less Fan Design. (2010). Excellence in engineering simulation. *Advantage* 4, 5–7.

[67] Nižetić, I., Fertalj, K., and Milašinović, B. (2007). *An Overview of Decision Support System Concepts*. Available at: http://old.foi.hr/CMS _home/znan_strucni_rad/konferencije/IIS/2007/papers/T06_01.pdf

[68] Gernhardt, G., Whitney, P., and Mandyck, J. (2016). *"The Future of Sustainable Aviation, Part 1 and 2."* Available at: http:// naturalleader.com/wp-content/themes/NaturalLeader2016/images/691 9-GreenAviationWhitePaper701.pdf

[69] Muslim, F. B., Demian, A., Ma, L., and Lavagno, L. (2016). *Energy-efficient FPGA Implementation of the k-Nearest Neighbours Algorithm Using OpenCL, Position Papers of the 2016 Federated Conference on Computer Science and Information Systems*. Available at: http:// www.ecoscale.eu/_docs/ecoscale_fedcsis16.pdf

[70] Shultza, J. M., and Garcia-Verab, M. P. (2013). Santiago de compostela train accident: disaster complexity and the santiago de compostela train derailment. *Dis. Health* 3, 1, 1–21.

[71] Philadelphia Train Accident. (2015). *Oversight of the Amtrak Accident in Philadelphia*. Available at: http://transportation.house.gov/uploaded files/2015-06-02-feinberg.pdf

[72] *Samsung to Disable Note 7 phones in recall effort for safety reasons, December 9, 2016*. Available at: https://phys.org/news/2016-1

[73] Blickstein, I., Boito, M., and Drezner, J. A. (2011). *Root Cause Analyses (note Added: Including trl Mismatch effects). Rand National Defense Research Institute*, Available at: http://www.rand.org/content/ dam/rand/pubs/monographs/2011/RAND_MG1171.1.pdf

[74] Leveson, N. (2004). A new accident model for engineering safer systems. *Saf. Sci.* 42, 237–270.

[75] UIC Safety Report. (2016). *UIC International Union of Railways (UIC) Issues Yearly Report 2016 on Railway Accidents in Europe*. Available at: http://safetydb.uic.org/IMG/pdf/SDB_2016.pdf

[76] Hall, A. D., and Fagen, R. E. (2017). *Definition of System*. http://www.isss.org/yearbook/1-C%20Hall%20&%20Fagen.pdf

[77] Mesarovic, M. D., Macko, D., and Takahara, Y. (1970). *Theory of Hierarchical Multilevel Systems*. New York, NY: Academic Press.

[78] Polderman, J. W., and Willems, J. C. (2014). *Introduction to the Mathematical Theory of Systems and Control*. Available at: http://wwwhome.math.utwente.nl/poldermanjw/onderwijs/DISC/math mod/book.pdf

[79] Jacobs, B. (2000). Object-oriented hybrid systems of co-algebras plus monoid actions. *Theor. Comp. Sci.* 239, 41–95.

[80] Rutten, J. (2011). *An Introduction to (Co)algebra and (Co)induction.* Available at: http://homepages.cwi.nl/janr/papers/files-of-papers/2011 _Jacobs_Rutten_new.pdf

[81] Rutten, J. J. M. M. (2000). Universal coalgebra: a theory of systems-fundamental study. *Theor. Comput. Sci.* 249, 3–80

[82] Mazzola, C. (2011). Becoming and the algebra of time, L&PS. *Logic Philos. Sci.* 9, 355–363.

[83] Akhundov, J. (2017). *Real -Time Systems 2017, section 2.2 Concept of Time.* Available at: https://osg.informatik.tu-chemnitz.de/lehre/ezs/rts-01-Basics-handout_en.pdf

[84] Lee, E. A. (2009). *Computing Needs Time, Technical Report No. UCB/EECS-2009-30.* Available at: http://www.eecs.berkeley.edu/Pubs/TechRpts/2009/EECS-2009-30.html

[85] Bowden, F. D. J. (2000). A brief survey and synthesis of the roles of time in petri nets. Math. Comput. Model. 31, 55–68.

[86] Bui, H. D. (2017). A distributed time triggered control for a feedback control system. *Int. Res. J. Eng. Technol.* 4, 1402–1407.

[87] Andrea Ceccarelli, Francesco Brancati, Bernhard Frömel, and Oliver Höftberger. (2016). "Time and resilient master clocks in cyber-physical systems," in *Cyber-Physical Systems of Systems, LNCS 10099*, ed. A. Bondavalli, S. Bouchenak, and H. Kopetz (Berlin: Springer) 165–185.

[88] Weiss, M. A., Li-Baboud, Y., Shrivastava, A., Khayatian, M., Derlery, P., Andrade, H. A., et al. (2016). "Time in Cyber-Physical Systems," in *Proceedings of the Association for Computing Machinery (ACM) International Conference on Hardware/Software-Codesign and Systems Synthesis (CODES+ISSS)*, New York, NY.

[89] Systems Engineering Body of Knowledge (SEBoK). (2017). *Systems of Systems (SoS)*. Available at: http://sebokwiki.org/wiki/Systems_of_Systems_(SoS)

[90] Dahmann, J., and Baldwin, K. (2008). Understanding the current state of US defense systems of systems and the implications for systems engineering. *Presented at IEEE Systems Conference, April 7–10, 2008, Montreal, Canada.*

[91] Maier, M. W. (1998). Architecting principles for systems-of-systems. *Syst. Eng.* 1, 267–284.

[92] DeLaurentis, D., and Crossley, W. (2005). A taxonomy-based perspective for system of systems design methods. *Paper 925, Presented at IEEE Conference on Systems, Man, and Cybernetics, October 10–12,*

2005, Waikoba, HI, Available at: https://www.tareasos.eu/docs/pb/SoA_V3.pdf

[93] Office of the Deputy Under Secretary of Defense for Acquisition and Technology, Systems and Software Engineering. (2008). *Systems Engineering Guide for Systems of Systems, Version 1.0.* Washington, DC: ODUSD(A&T)SSE, 2008. Available at: http://www.acq.osd.mil/se/docs/SE-Guide-for-SoS.pdf

[94] Burns, A., and Davis, R. I. (2017). *Mixed Criticality Systems*. Available at: https://www-users.cs.york.ac.uk/burns/review.pdf

[95] European Mixed-Criticality Cluster. (2016). *Tackling Future Challenges in the Design and Development of Mixed-Criticality Multicore Systems.* Available at: https://www.uni-siegen.de/dreams/community/cluster/eu-mixed-criticality-cluster_factsheet.pdf

[96] Crespo, A., Alonso, A., de la Puente, M. J. A., and Balbastre, P. (2014). *Mixed Criticality in Control Systems.* Cape Town: 19th IFAC World Congress, 12262–12271.

[97] Baruah, S. K., Burns, A., and Davis, R. I. (2011). *Response-Time Analysis for Mixed Criticality Systems.* Available at: https://www-users.cs.york.ac.uk/burns/RTSS.pdf

[98] Ernst, R. (2010). *Certification of Trusted MpSoC Platforms (RECOMP Project 2010).* Available at: http://www.mpsoc-forum.org/previous/2010/slides/6-Ernst.pdf

[99] Herkersdorf, A., and Paulitsch, M. (2013). Multicore enablement for embedded and cyber physical systems, dagstuhl seminar 2013-13052. *Dagstuhl Rep.* 3, 149–182, eds A. Herkersdorf and M. Paulitsch. Available at: http://drops.dagstuhl.de/opus/volltexte/2013/4015/pdf/dagrep_v003_i001_p149_s13052.pdf

[100] Lee, E. A. (2015). "Architectural support for cyber-physical systems, keynote ASPLOS 2015," in *Proceedings of the Twentieth International Conference on Architectural Support for Programming Languages and Operating Systems*, Istanbul.

[101] Eidson, Lee, E. A., Matic, S., Sheshia, S., and Zou, J. (2012). Distributed real-time software for CPS. *IEEE Proc.*

[102] Lee, E. A. (2015). The past, present and future of cyber-physical systems: a focus on models. *Sensors* 15, 4837–4869.

[103] Lee, E. A. (2006). "Cyber-physical systems-are computing foundations adequate?," *Position Paper for NSF Workshop on Cyber-Physical Systems: Research Motivation, Techniques and Roadmap*, Austin, TX.

[104] Lee, E. A. (2016). Fundamental limits of cyber-physical systems modeling. *ACM Trans. Cyber-Phys. Syst.* 1. doi: 10.1145/2912149

[105] Carayannis, E. G., Campbell, D. F. J., and Rehman, S, S. (2016). Mode 3 knowledge production: systems and systems theory, clusters and networks. *J. Innovat. Entrepreneurship* 5:17.

[106] Zouaghi, I., and Spalanzani, A. (2009). "Supply chains, Ago-antagonistic systems through co-opetition game theory lens," in *CAHIER DE RECHERCHE n2009-13 E5, International Logistics and Supply Chain Congress' 2009*, Istanbul. Available at: https://halshs.archives-ouvertes.fr/halshs-00540290/document

[107] Olfati-Saber, R., Alex Fax, J., and Murray, R. M. (2007). Consensus and cooperation in networked multi-agent systems. *Proc. IEEE* 95, 215–233.

[108] Burgos, A. C., and Polani, D. (2016). *Cooperation and Antagonism in Information Exchange in a Growth Scenario with Two Species.* Available at: https://arxiv.org/pdf/1505.00950.pdf

[109] Freilich, S., Zarecki, R., and Eilam, O. (2011). Competitive and cooperative metabolic interactions in bacterial communities. *Nat. Commun.* 2:589.

[110] Chae, E., Tran, D. T. N., and Weigel, D. (2016). Cooperation and conflict in the plant immune system. *PLOS Pathog.* 12:e1005452.

[111] Meng, D., Jia, Y., and Du, J. (2015). Finite-time consensus for multiagent systems with cooperative and antagonistic interactions. *IEEE Trans. Neural Netw. Learn. Syst.* 27, 762–770. doi: 10.1109/TNNLS.2015.2424225

[112] Loerakker, B., Bault, N., Hoyer, M., and van Winden, F. (2016). *Asymmetry in the Development of Cooperative and Antagonistic Relationships. A Model-Based Analysis of a Fragile Public Good Game Experiment, Repor, Center for Research in Experimental Economics and political Decision-making (CREED), Amsterdam School of Economics, University of Amsterdam, 23-12-2016.* Available at: http://www.creedexperiment.nl/creed/pdffiles/AsymmetryDevelopmentTies.pdf

[113] Vassea, M., Noblea, R J., and Akhmetzhanov, A. R. (2017). Antibiotic stress selects against cooperation in the pathogenic bacterium *Pseudomonas aeruginosa. Proc. Natl. Acad. Sci. U.S.A.* 114, 546–551. doi: 10.1073/pnas.1612522114

[114] Shehory, O., and Kraus, S. (1996). "Cooperative goal-satisfaction without communication in large-scale agent-systems," in *ECAI 1996*

12th European Conference on Artificial Intelligence, ed. W. Wahlster (Budapest), 544–548.

[115] Sifakis, J. (2014). *Rigorous System Design, RTSS, Rome*. Available at: http://2014.rtss.org/wp-content/uploads/2014/12/RTSS2014-sifakis-distr.pdf

[116] Mavridou, A. (2016). *Modelling Architecture Styles*, Ph.D. thesis, École Polytechnique Fédérale, Lausanne, 7324.

[117] Konnov, I., Kotek, T., Wang, Q., Veith, H., Bliudze, S., and Sifakis, J. (2016). "Parameterized systems in BIP: design and model checking," in *Proceedings of the 27th International Conference on Concurrency Theory (CONCUR 2016)*, eds J. Desharnais and R. Jagadeesan (Dagstuhl: Schloss Dagstuhl – Leibniz-Zentrum für Informatik GmbH), doi: 10.4230/LIPIcs.CONCUR.2016.30

[118] Mavridou, A., Stachtiari, E., Bliudze, S., Ivanov, A., Katsaros, P., and Sifakis, J. (2016). "Architecture-based design: a satellite on-board software case study," in *Proceedings of the 13th International Conference on Formal Aspects of Component Software (FACS 2016)*, eds O. Kouchnarenko and R. Khosravi (Berlin: Springer). Available at: https://indico.esa.int/indico/event/146/contribution/21/material/1/0.pdf

[119] Bliudze, S., Mavridou, A., Szymanek, R., and Zolotukhina, A. (2017) Exogenous coordination of concurrent software components with Java-BIP. *Softw. Pract. Exper.* doi: 10.1002/spe.2495

[120] Falcone, Y., Jaber, M., Nguyen, T-H., Bozga, M., and Bensalem, S. (2011). *Runtime Verification of Component-based Systems*, Available at: https://www.irisa.fr/vertecs/Publis/Ps/sefm11.pdf

[121] Sangiovanni-Vincentelli, A., Damm, W., and Passerone, R. (2012). Taming Dr. frankenstein: contract-based design for cyber-physical systems. *Eur. J. Control.* 18, 217–238.

[122] Benveniste, A., Caillaud, B., and Passerone, R. (2007). *A generic model of contracts for embedded systems. Research report 6214, IRISA/INRIA Rennes*. Available at: https://hal.inria.fr/inria-00153477/document

[123] Staron, M. (2011). Software Complexity Metrics in general and in context of ISO 26262 software verification requirements. Available at: http://safety.addalot.se/upload/2016/PDF/1-9_Staron_complexity.pdf

[124] Khan, A. A., Mahmood, A., Amralla, S. M., and Mirza, T. H. (2016). Performance comparison of software complexity metrics. *Int. J. Com. Net. Tech.* 4, 106–174.

[125] European Research Council. (2017). *REF European Research Council Classification of S&T domains (ERC)*. Available at: https://erc.

europa.eu/sites/default/files/document/file/erc%20peer%20review%20
evaluation%20panels.pdf

[126] DFG Classification of scientific disciplines (2016–2019). *DFG Classification of Scientific Disciplines*. Available at: http://www.dfg.de/down
load/pdf/dfg_im_profil/gremien/fachkollegien/amtsperiode_2016_2019
/fachsystematik_2016-2019_en_grafik.pdf

[127] Department of Economic and Social Affairs. (2008). *International Standard Industrial Classification of All Economic Activities Revision 4*. New York, NY: United Nations. Available at:
https://unstats.un.org/unsd/publication/seriesM/seriesm_4rev4e.pdf

[128] OECD calculations based on IPO. (2014). *Eight Great Technologies: the Patent Landscapes, United Kingdom, and STI Micro-data
Lab: Intellectual Property database*. Available at: www.gov.uk/govern
ment/publications/eight_great_technologies_the_patent_landscapes

[129] OECD. (2015). *Frascati Manual 2015*. Available at: http://www.oecd.
org/sti/inno/38235147.pdf

[130] Iyengar, S., and Brooks, R. (2005). *Distributed Sensor Networks,
C&H*. Boca Raton, FL: CRC Book.

[131] Cardenas, A. A., Amin, S., Sinopoli, B., Giani, A., Perrig, A., and
Sastry, S. (2009). *Challenges for Securing Cyber Physical Systems*.
https://chess.eecs.berkeley.edu/pubs/601/cps-security-challenges.pdf

[132] Humayed, A., Lin, J., Li, F., and Luo, B. (2017). *Cyber-Physical Systems Security: A Survey*. Available at: https://arxiv.org/pdf/1701.0452
5.pdf

[133] Scientific Foresight Study. (2016). *Scientific Foresight study Ethical
Aspects of Cyber-Physical Systems*. Available at: http://www.europarl.
europa.eu/RegData/etudes/STUD/2016/563501/EPRS_STU(2016)563
501_EN.pdf

[134] Geisberger, E., and Broy, M. (2012). *Integrierte Forschungsagenda
Cyber Physical Systems", Acatech Studie*, Berlin: Springer. doi
10.1007/978-3-642-29099

[135] Holmes, D. S., and Courts, S. S. (2016). *Resolution and Accuracy
of Cryogenic Temperature Measurements*, Westerville: Lake Shore
Cryotronics, Inc.

[136] Dempsey, P. J., and Fabik, R. H. (1993). *Cryogenic Temperature Measurements*, Diodes: Lake Shore Cryotronics. Available at:
www.lakeshore.com

[137] Fraunhofer IOSB (2017). Glossar Industrie 4.0 des Fachausschuss VDI/VDE-GMA 7.21 Industrie 4.0, Verabschiedete Begriffsdefinitionen, Stand 9.11.15, Miriam Schleipen, Fraunhofer IOSB, Available at: http://www.iosb.fraunhofer.de/?BegriffeI40

[138] Wang, Z., Liu, Z., and Zheng, C., (2016). Qualitative *Analysis and Control of Complex Neural Networks with Delays, Studies in Systems, Decision and Control 34.* Berlin: Springer-Verlag, doi: 10.1007/978-3-662-47484-6_2

[139] Giese, H., Rumpe, B., SchÃtz, B., and Sztipanovits, B. (2012). *Science and Engineering of Cyber-Physical Systems, Report from Dagstuhl Seminar 11441.* Available at: http://drops.dagstuhl.de/opus/volltexte/2012/3375/pdf/dagrep_v001_i011_p001_s11441.pdf

[140] Maheshwari, A., Kenley, C. R., and DeLaurentis, D. A. (2015). "Creating Executable Agent-Based Models Using SysML," in *Proceedings of the 25th Annual INCOSE International Symposium (IS2015),* Bellevue, WA. Available at: https://www.researchgate.net/profile/Apo orv_Maheshwari/publication/283967514_Creating_Executable_Agent-Based_Models_Using_SysML/links/570d15f008ae2b772e42cd4f.pdf? origin=publication_detail

[141] Ratlif, L. J. (2015). *Incentivizing Efficiency in Societal-Scale Cyber-Physical Systems,* Ph.D. thesis, UCB m Berkeley, Berkeley, CA.

[142] Lee, J., Bagheri, B., and Kao, H.-A. (2015). A cyber-physical systems architecture for Industry 4.0-based manufacturing systems. *Manufact. Lett.* 3, 18–23. doi: 10.1016/j.mfglet.2014.12.001

[143] Mario, H., Tobias, P., and Boris, O. (2014). *Design Principles for Industry 4.0 Scenarios: A Literature Review.* St. Gallen: Technische Universität Dortmund.

[144] László Monostori, L. (2014). "Cyber-physical production systems: roots, expectations and R&D challenges," in *Proceedings of the 47th CIRP Conference on Manufacturing Systems* (Amsterdam: Elsevier), doi: 10.1016/j.procir.2014.03.115

[145] Samad, T. (2016). Control systems and the internet of things. *IEEE Control Syst. Mag.* 13–16.

[146] Grosu, R. (2015). *Cyber-physical Systems,* Vienna: Vienna Technical University.

[147] Bartocci, E., Hoeftberger, O., and Grosu, R. (2004). *Cyber-Physical Systems: Theoretical and Practical Aspects, ERCOM NEWS 97,* Available at: http://ti.tiwien.ac.at/cps

[148] Poovendran, R. (2010). *Cyber-Physical Systems: Close Encounters Between Two Parallel Worlds.* Seattle, WA: University of Washington. doi 10.1109/JPROC.2010.2050377

[149] Leveson, N. (2013). The drawbacks of using the term "System of systems." *Biomed. Instrument. Technol.* 47, 115–118.

[150] Pronios, N. B. (2015). *CPS and other "smart" systems: A review, Innovate UK, H2020 funding for Smart Systems and ICT for Factories of the Future London.* Available: https://connect.innovateuk.org/docume nts/407758/30061168/Systems+overview_London_05_11_15_f.pdf/5cc 9d758-0785-4725-915c-4319a24179af

[151] Samad, T. (2014). *The Web of Things and Cyber-Physical Systems: Closing the Loop, Report,* Morris Plains, NJ: Honeywell.

[152] VDI/VDE/GMA (Prof. Stefan Kowalewski). (2013). *Cyber-Physical Systems: Chancen und Nutzen aus Sicht der Automation, Thesen und Handlungen.* Düsseldorf: Verein Deutscher Ingenieure.

[153] Kowalewski, I. S., Rumpe, B., and Stollenwerk, A. (2014). *Cyber-Physical Systems, Eine Herausforderung für die Automatisierungstechnik?, RWTH Aachen.* Available at: http://www.se-rwth.de/publications/Cyber-Physical-Systems-eine-Herausforderung-an-die-Automatisierungstechnik.pdf

[154] Novikov, D. A. (2016). *Cybernetics – From Past to Future,* Heidelberg: Springer.

[155] Augusto, J. C. (2006). *Ambient Intelligence, Basic Concepts and Applications.* Available at: http://hssc.sla.mdx.ac.uk/staffpages/juanau gusto/jca_KN_ICSOFT.pdf

[156] Werbos, P. J. (2015). "Computational intelligence from AI to BI to NI," in *Proceedings of the SPIE 9496, Independent Component Analyses, Compressive Sampling, Large Data Analyses (LDA), Neural Networks, Biosystems, and Nanoengineering XIII, 94960R (May 20, 2015),* Baltimore, MD. doi: 10.1117/12.2191520

[157] Stankovic, J. A., Lee, I., Mok, A., and Rajkumar, R. (2005). *Carnegie Mellon University: "Physical Computing," IEEE Computer 2005,* Available at: http://www.cis.upenn.edu/lee/10cis541/papers/05nov-computer-final.pdf

[158] Alur, R., D'Antoni, L., Deshmukh, J. V., Ragothaman, M., and Yuan, Y. (2012). *Regular Functions, Cost Register Automata, and Generalized Min-Cost Problems.* Available at: http://www.cis.upenn.edu/ alur/rca12.pdf

[159] Alur, R., D'Antoni, L., Deshmukh, J. V., Ragothaman, M., and Yuan, Y. (2012). *Regular Functions and Cost Register Automata*. Available at: https://www.cis.upenn.edu/ alur/Lics13reg.pdf

[160] Alur, R., Fisman, D., and Raghothaman, M. (2016). *Regular Programming for Quantitative Properties of Data Streams*. Available at: https://www.cis.upenn.edu/alur/Esop16.pdf

[161] Alur, R., and Tripakis, S. (2015). *Automatic Synthesis of Distributed Protocols, SIGACT 20127*. Available at: https://www.cis.upenn.edu/ alur/Sigact17.pdf

[162] Matsikoudis, E., and Lee, E. A. (2014). *Generalized Ultrametric Semilattices of Linear Signals Technical Report No. UCB/EECS-2014-7*. Available at: http://www.eecs.berkeley.edu/Pubs/TechRpts/2014/EECS-2014-7.html

[163] Matsikoudis, E., and Lee, E. A. (2015). The fixed-point theory of strictly causal functions. *Theor. Comp. Sci.* 574, 39–77.

[164] Xu, D. P., Jahanchahi, C., Took, C. C., and Mandic, D. P. (2014). *Quaternion Derivatives: The GHR Calculus, Technical Report, TR-Quat-GHR-Calc-08-01-14/1 Imperial College London*. Available at: http://www.commsp.ee.ic.ac.uk/mandic/Technical_Report/TR_ICL_Quaternion-GHR_Jan_2014.pdf

[165] Bihan, N. L. (2015). *The Geometry of Proper Quaternion Random Variables*. Available at: https://arxiv.org/pdf/1505.06182.pdf

[166] Wang, M., Took, C. C., and Mandic, D. P. (2011). "A class of fast quaternion valued variable stepsize stochastic gradient learning algorithms for vector sensor processes," in *Proceedings of International Joint Conference on Neural Networks*, San Jose, CA. Available at: http://www.commsp.ee.ic.ac.uk/mandic/research/Quaternion_Stuff/Quaternion_Adaptive_GASS_IJCNN_2011.pdf

[167] Alfsmann, D., Göckler, H. G., Sangwine, S. J., and Ell, T. A. (2007). "Hypercomplex algebras in DSP benefits and drawbacks," in *15th European Signal Processing Conference (EUSIPCO 2007), Poznan, Poland, September 3–7*. Available at: http://www.eurasip.org/Proceedings/Eusipco/Eusipco2007/Papers/c2l-b01.pdf

[168] Molodtsov, D. A. (1999). Soft set theory"first results. *Comput. Math. Appl.* 37, 19–31. doi: 10.1016/S0898-1221(99)00056-5

[169] Zhang, X. (2014). On interval soft sets with applications. *Int. J. Comput. Intell. Syst.* 7, 186–196.

[170] Zhan, J., Ali, M. I., and Mehmood, N. (2017). *On a Novel Uncertain Soft Set Model: Z-Soft Fuzzy Rough Set Model and Corresponding*

Decision Making Methods. Available at: https://www.researchgate.net/
publication/315930689_On_a_novel_uncertain_soft_set_model_Z_-soft_
fuzzy_rough_set_model_and_corresponding_decision_making_methods

[171] Wei, B., Xie, Q., Meng, Y., and Zou, Y. (2017). *Fuzzy Geography Mark-up Language (GML) Modelling Based on Vague Soft Sets. ISPRS Int. J. Geo-Inf*. 6:10. doi: 10.3390/ijgi6010010

[172] Alkhazaleh, S. (2015). The multi-interval-valued fuzzy soft set with application in decision making. *Appl. Math.* 6, 1250–1262.

[173] Brunsch, T., Hardouin, L., Maia, C. A., and Raisch, J. (2017). Duality and interval analysis over idempotent semirings. *Lin. Algebra Appl.*

[174] Godsil, C., and Zhan, H. (2017). Discrete-time quantum walks and graph structures. Available at: https://arxiv.org/pdf/1701.04474.pdf

[175] Krebbers, R., Timany, A., and Birkedal, L. (2017). "Interactive proofs in higher-order concurrent separation logic," in *Proceedings of the 44th ACM SIGPLAN Symposium on Principles of Programming Languages* (New York, NY: ACM).

[176] Legay, A., and Margaria, T (eds) (2017). *Tools and Algorithms for the Construction and Analysis of Systems*. Berlin: Springer.

[177] Gainutdinova, A., and Yakaryilmaz, A. (2016). *Non-deterministic unitary OBDDs*. Available at: https://arxiv.org/pdf/1612.07015.pdf

[178] Aiguier, M., and Kanso, B. (2014). A logic for complex computing systems: properties preservation along integration and abstraction. *Sci. Ann. Comput. Sci.* 24, 1–46. doi: 10.7561/SACS.2014.1.1

[179] Wolff, K. E. (2000). *Towards a Conceptual System Theory, and "A Conceptual View on Rough Set Theory."* Jaipur: Rajasthan State Road Transport Corporation.

[180] Ponsard, C., and De Landtsheer, R. (2016). Comparison of the AADL and Event-B model-based tool chain for designing embedded systems.

[181] Kottenstette, N., Hall, J., Koutsoukos, X., Sztipanovits, J., and Antsaklis, P. (2013). Design of networked control systems using passivity. *IEEE Trans. Control Syst. Technol.* 21, 649–665.

[182] Antsaklis, P. J., Goodwine, B., Gupta, V., McCourt, M. J., Wang, Y., Wu, P., et al. (2013). Control of cyberphysical systems using passivity and dissipativity based methods. *Eur. J. Control* 19, 379–388.

[183] Schaft, A. J. V. D., and Maschke, B. M. (2013). Port-hamiltonian systems on graphs. *SIAM J. Control Optimiz.* 51, 906–937.

[184] Kaburlasos, V., and Petridis, V. (2000). *Learning and Decision Making in the Framework of Fuzzy Lattices*. Available at: https://www.research

gate.net/publication/2805473_Learning_and_Decision-Making_in_the_
Framework_of_Fuzzy_Lattices

[185] 5G Vision. (2015). *The 5G Infrastructure Public Private Partnership: the next generation of communication networks and services, European Commission 2015.* Available at: https://5g-ppp.eu/wp-content/uploads/2015/02/5G-Vision-Brochure-v1.pdf

[186] https://5g-ppp.eu/white-papers/, https://5g-ppp.eu/wp-content/uploads/2014/02/5G-PPP-5G-Architecture-WP-July-2016.pdf

[187] Mitra, R. N., and Agrawal, D. P. (2015). *5G mobile technology: a survey.* ICT Exp. 1, 132–137.

[188] Osseiran, A., Monserrat, J. F., and Marsch, P. (eds). (2016). *5G Mobile and Wireless Communications Technology.* Cambridge: Cambridge University Press.

[189] ETSI (2016). *5G from Myth to Reality.* Available at: https://docbox.etsi.org/Workshop/2016/20160421_5G_FROM_MYTH_TO_REALITY/SESSION_D_ULTRA_RELIABLE_LOW_LATENCY_NWKS/INDUSTRIE_4_0_ANY_NEED_FOR_5G_WEGENER_ZVEI.pdf

[190] "INDUSTRIE 4.0" Any Need for 5G? Prof. Dr. Dieter Wegener, Speaker ZVEI-management-team "Industrie 4.0", RAMI 4.0 Reference-Architecture-Model "Industrie 4.0".

[191] Barot, V., Henshaw, M., Siemieniuch, C., Sinclair, M., Lim, S. L., and Henson, S. (2013). *Mo Jamshidi, Daniel DeLaurentis: SoS State of the Art Report of the joint EU-US project T-AREA-SoS,* Loughborough: Loughborough University. Available at: https://www.tareasos.eu/docs/pb/SoA_V3.pdf

[192] AMADEOS EC Project (D1.1 – SoSs, commonalities and requirements Rev. 3.11): (2014). *Architecture for Multi-criticality Agile Dependable Evolutionary Open System-of-Systems.* Available at: http://amadeos-project.eu/wp-content/uploads/2014/10/AMADEOS_WP1_D1.1_v3.11final-version.pdf

[193] Siljak, D. (2007). *Large Scale Dynamic Systems, Stability & Structure,* Dover: North Holland.

[194] Astrom,K.-J., Albertos, P., Blanke, M., Isidori, A., Schaufelberger, W., Sanz, R. et al. (2001). *Control of Complex Systems.* Berlin: Springer 2001.

[195] Koutsoukos, X., and Antsaklis, P. (2006). Science for Cyber-Physical Systems Available at: http://ieeecss.org/sites/ieeecss.org/files/CSSIoCT2Update/IoCT2-RC-Koutsoukos-1.pdf

[196] Karkanias, N. (2016). *"Complex Systems and Control."* London: City University.

[197] Engell, S. (2013). *State of the Art and Future Directions in Management and Control of Physically Coupled Systems of Systems*. Available at: http://www.dymasos.eu/wp-content/uploads/2016/10/DYMASOS_D5.5_final.pdf

[198] Internet of Things (IoT)/Cyber-physical Systems (CPS) Report (2016). *PICASO project*. Available at: http://www.picasso-project.eu/wp-content/uploads/2016/06/Report-from-IoT-CPS-Expert-Group_20 16-05-20-pm-1.pdf

[199] St. Gallen Symposium. (2008). *"Transform your company's decision-making strategy," Voices of the Leaders of Tomorrow*. Available at: http://www.gfk-verein.org

[200] Makarov, I. M., Vonogradskaya, T. M., Rubchinsky, A. A., and Sokolov, V. B. (1982/1987). *The Theory of Choice and Decision Making*. Moskow.

[201] Fradkov, A.L. *Cybernetical Physics: Principles and Examples*. Nauka, St. Petersburg, 2003, p. 208 (in Russian).

[202] Hu, T.-W., and Kaneko, M (2014). *Undecidability arising from Prediction/Decision Making in an Interdependent Situation*. Available at: http://www.saet.uiowa.edu/images/th2014aug10.pdf

[203] Tripakis. S. (2004). Undecidable problems of decentralized observation and control on regular languages. *Inform. Proc. Lett.* 90, 21–28.

[204] Platzer, A. (2016) "Logic & proofs for cyber-physical systems," in *Automated Reasoning. IJCAR 2016. Lecture Notes in Computer Science*, Vol. 9706, eds N. Olivetti and A. Tiwari (Cham: Springer).

[205] Betrane, J., Cousot, P., Cousot, R., Feret, J., Mauborgne, K., Mine, A., et al. (2010). *Static Analysis and Verification of Aerospace Software by Abstract Interpretation*. Atlanta: Arizona Interscholastic Athletic Administrators Association, 1–36.

[206] Mitsch, S., and Platzer, A. (2014). ModelPlex: verified runtime validation of verified cyber-physical system models. *Int. Conf. Runtime Verifcation* 8734, 199–214.

[207] Ren, W., and Beard, R. W. (2008). *Distributed Consensus in Multi-Vehicle Cooperative Control Theory and Applications*. Berlin: Springer.

[208] Schirner, G., Erdogmus, D., and Chowdhury, K. (2013). *The Future of Human in-the-Loop Cyber-Physical Systems*. Available at:

http://www1.ece.neu.edu/schirner/cv/J5_Dagstuhl_Computer_Jan_13_ HiLCPS.pdf

[209] Stankovic, J. A., Mike Liang, C.-J., and Lin, S. (2013). *Cyber Physical System Challenges for Human-in-the-Loop Control.* Available at: https://www.microsoft.com/en-us/research/wp-content/uploads/2016/ 02/feedback13.pdf

[210] Seshia, S. A. (2015). *Verification by, for, and of Humans: Formal Methods for Cyber-Physical Systems and Beyond.* Available at: https:// people.eecs.berkeley.edu/ sseshia/talks/Seshia-UIUCColloquium.pdf

[211] SAE. (2016). *SAE Automated Driving Standard 2016.* Available at: https://www.sae.org/misc/pdfs/automated_driving.pdf

[212] Tamiz, M., Jones, D., and Romero, C. (1998). Goal programming for decision making: An overview of the current state-of-the-art. *Eur. J. Operat. Res.* 111, 569–581.

[213] Marler, R. T., and Arora, J. S. (2004). Survey of multi-objective optimization methods for engineering. *Struct. Multidisc. Optim.* 26, 369–395. doi: 10.1007/s00158-003-0368-6

[214] Mendes, L. F., Ding, B., and Han, J. (2011). Stream sequential pattern mining with precise error bounds. *Anal. Chem.* 83, 5631–5638. doi: 10.1021/ac200740w

[215] Zhang, J., Gonzalez, E., Hestilow, T., Haskins, W., and Huang, Y. (2009). Review of Peak Detection Algorithms in Liquid-Chromatography-Mass Spectrometry. *Curr. Genom.* 10, 388–401.

[216] Abbas, H., Rodionova, A., Bartocci, E., Smolka, S. A., and Grosu, R. (2016) *Regular Expressions for Irregular Rhythms.* Available at: https://arxiv.org/pdf/1612.07770.pdf

[217] Heylighen, F., and Joslyn, C. (2001). "Cybernetics and second-order cybernetics," in *Encyclopedia of Physical Science & Technology*, 3rd Edn, ed. R. A. Meyers (New York, NY: Academic Press).

[218] Soroudi, A., and Amraee, T. (2013). Decision making under uncertainty in energy systems: State of the art. *Renew. Sustain. Energy Rev.* 28, 376–384.

[219] Feng, M.-L., and Tan, Y.-P. (2004). "Adaptive binarization method for document image analysis," in *Proceedings of the IEEE International Conference on Multimedia and Expo (ICME).*

[220] Wong, P. Y. H., and Gibbons, J. (2007). "A process-algebraic approach to workflow specification and refinement," in *SC 2007, LNCS 4829*, eds M. Lumpe and W. Vanderperren (Berlin: Springer), 51–65.

[221] Siozios, K., Danassis, P., Zompakis, N., Korkas, C., Kosmatopou-
      los, E., and Soudris, D. (2015). "Supporting decision making for
      large-scale iots: trading accuracy with computational complexity,"
      in *Components and Services for IoT Platforms*, Chap. 12, eds G.
      Keramidas, N. Voros, and M. Hübner (Berlin: Springer).
[222] Kashevnik, A. (2015). *Ontology-Based Robots Self-Organization
      in Cyber-Physical Systems*. Petrozavodsk: AMICT. Available at:
      http://cs.petrsu.ru/fdpw/2015/presentation/kashevnik.pdf
[223] Brunsch, T., Raisch, J., and Hardouin, L. (2012). Modeling and control
      of high-throughput screening systems. *Control Eng. Pract.* 20, 14–23.
[224] GAMS–CyBio Scheduler. (2015). *CyBio Scheduler "Scheduling Soft-
      ware for High Throughput Screening, A GAMS Application in the
      Pharmaceutical Industry*. Available at: https://www.gams.com/case-
      studies/cybio-scheduler/
[225] Onkena, A.-K., Badera, A., and Trachta, K. (2016). Logistical control
      of flexible processes in high-throughput systems by order release and
      sequence planning. *Procedia CIRP* 52, 245–250.
[226] Schuck, B. W., MacArthur, R., and Inglese, J. (2017). Quantitative
      high-throughput screening using a coincidence reporter biocircuit.
      Curr. Protoc. Neurosci. 79, 5.32.1–5.32.27. doi: 10.1002/cpns.27
[227] Sifakis, J. (2012). Rigorous system design, foundations and trends.
      Electron. Design Automat. 6, 293–362.
[228] Saadatmand, M. (2012). *Satisfying Non-Functional Requirements
      in Model-Driven Development of Real-Time Embedded Systems*.
      Västerås: Mälardalen University.
[229] Hochmüller, E., and Dobrovnik, M. (2006). *Identifying Types of Extra-
      Functional Requirements in the Context of Business Process Support
      Systems*. Available at: http://ceur-ws.org/Vol-75/files/REBPS_06.pdf
[230] Poort, E. R., Martens, N., van de Weerd, I., and van Vliet, H. (2012).
      "How architects see non-functional requirements," in *REFSQ 2012,
      LNCS 7195*, eds B. Regnell and D. Damian (Berlin: Springer), 37–51.
[231] Blanchette, S. Jr., Albert, C., and Garcia-Miller, S. (2010). *Beyond
      Technology Readiness Levels for Software:, U.S. Army Work-
      shop Report, TECHNICAL REPORT, CMU/SEI-2010-TR-044, ESC-
      TR-2010-109*. Available at: http://resources.sei.cmu.edu/library/asset-
      view.cfm?assetid=9689
[232] Sauser, B., Ramirez-Marquez, J. E., Magnaye, R., and Tan, W. (2009).
      Systems approach to expanding the technology, readiness level within
      defense acquisition. *Int. J. Def. Acquisit. Manag.* 1, 39–58.

[233] Sauser, B. (2012). *System Maturity and Architecture Assessment, Methods, Processes, and Tools Stevens Institute of Technology, Technical Report SERC-2012-TR-027-1.* Available at: http://www.dtic.mil/docs/citations/ADA582064

[234] Olechowski, A., Eppinger, S. D., and Joglekar, N. (2015). "Technology readiness levels at 40' and opportunities," in *Proceedings of the Portland International Conference on Management of Engineering and Technology (PICMET)*, Portland, OR.

[235] Jullien, C. (2014). *Considerations for an Innovation Readiness Level along with the Technology and Manufacturing Readiness Level indicators IEA Committee on Energy Research and Technology.* Paris: Modelling and Analyses in R&D Priority-Setting and Innovation.

[236] Azizian, N., Sarkani, S., and Mazzuchi, T. (2009). "A comprehensive review and analysis of maturity assessment approaches for improved decision support to achieve efficient defence acquisition," in *Proceedings of the World Congress on Engineering and Computer Science 2009, Vol II, (WCECS 2009)*, San Francisco, CA.

[237] Ross, S. (2016). Application of system and integration readiness levels. *ARJ* 23, 248–273.

[238] Abrial, J.-R. (2010). *Modelling in Event-B: System and Software Engineering.* Cambridge: Cambridge University Press.

[239] Jakovljevic, M. (2015). *Robust Embedded Computing for Advanced Integrated Architectures.* Vienna: TTtech.

[240] Reed, G. M., and Roscoe, A.W. (1988). A timed model for communicating sequential processes. *Theor. Comput. Sci.* 58, 249–261.

[241] Alkis Konstantellos: "A short overview of control in European R&D programmes (1983–2013): From local loop designs, through networked and coordinated control, to stochastic, large scale and real-time optimization systems", in "The Path of Control", Edited by Alessandro Astolfi, Special Issue, European Journal of Control, Volume 19, Issue 5, pp. 351–357 (September 2013), https://doi.org/10.1016/j.ejcon.2013.06.002

[242] Roger Penrose: The Road to Reality: A Complete Guide to the Laws of the Universe, Jonathon Cape Random House, London (2004).

[243] Danielle C. Tarraf (Ed.): Control of Cyber-Physical Systems, Workshop held at Johns Hopkins University, March 2013, https://doi.org/10.1007/978-3-319-01159-2

[244] Wenchao Li, Dorsa Sadigh, S. Shankar Sastry and Sanjit A. Seshia*/: Synthesis for Human-in-the-Loop Control Systems. *International*

Conference on Tools and Algorithms for the Construction and Analysis of Systems (TACAS), April 2014.

[245] Jyri Rajamäki: Resilient Sociotechnical, Cyber-Physical, Software-Intensive Systems of Systems ISSN: 2367-8895 1 Volume 2, 2017.

[246] Nima Asadi, Mehrdad Saadatmand, Mikael Sjodin: Run-Time Monitoring of Timing Constraints: A Survey of Methods and Tools, ICSEA 2013: The Eighth International Conference on Software Engineering Advances, http://www.es.mdh.se/pdf_publications/3012.pdf

[247] R. Obermaisser H Kopetz (eds), GENESYS: A Candidate for an ARTEMIS Cross-Domain Reference Architecture for Embedded Systems, Oct 2009, http://citeseerx.ist.psu.edu/viewdoc/download?doi= 10.1.1.149.388&rep=rep1&type=pdf

[248] HIPEAC Vision, https://www.hipeac.net/publications/vision/

[249] Horizon 2020, Work Programme 2016 –2017, European Commission, Brussels, http://ec.europa.eu/research/participants/data/ref/h2020/wp/ 2016_2017/main/h2020-wp1617-intro_en.pdf

[250] European Information and Communications Technologies Observatory (EITO), https://www.eito.com/

[251] Tomas Bures, Danny Weyns, Christian Berger, Stefan Biffl, Marian Daun, Thomas Gabor, David Garla, Ilias Gerostathopoulos, Christine Julien, Filip Krikava, Richard Mordinyi, Nikos Pronios: Software Engineering for Smart Cyber-Physical Systems – Towards a Research Agenda, Report on the First International Workshop on Software Engineering for Smart CPS.

[252] ISA100, Industrial Wireless Protocol, http://isa100wci.org/Documents /PDF/The-Technology-Behind-ISA100-11a-v-3_pptx0

[253] Daniel K. Molzahn, Florian Dorfler, Henrik Sandberg, Steven H. Low, Sambuddha Chakrabarti Ross Baldick: A Survey of Distributed Optimization and Control Algorithms for Electric Power Systems, http://www.ieor.berkeley.edu/~lavaei/Survey_Dis_2017.pdf

[254] Attila Kozma: Distributed Optimisation Methods for Large Scale Optimal Control, PhD thesis, November 2013, KU Leuven.

[255] Federica Garin, Luca Schenato: A survey on distributed Estimation and Control Applications Using Linear Consensus Algorithms, http://necs.inrialpes.fr/people/garin/WIDEbook_GarinSchenato.pdf

[256] Marios M. Polycarpou, Yanli Yang and Kevin M. Passino, Cooperative Control of Distributed Multi-Agent Systems, IEEE Control Systems Magazine (June 2001).

[257] Carlos Canudas de Wit, On-line partitioning algorithms for evolutionary scale-free Networks, http://www.gipsa-lab.grenoble-inp.fr/
~carlos.canudas-de-wit/ERC_positions/PhD2-ERC-Scale-FreeBack_
Partitions.pdf [ERC Grant for 2016–2021, European Commission].

[258] Boucherie, R. J., & van Dijk, N. M. (Eds.) (2017). *Markov Decision
Processes in Practice*. (International Series in Operations Research &
Management Science; Vol. 248). Cham: Springer International Pub-
lishing. DOI: 10.1007/978-3-319-47766-4

[259] David Dillenberger, Andrew Postlewaite, Kareen Rozen: Optimism
and Pessimism with Expected Utility, March 2016, https://web.sas.upe
nn.edu/ddill/files/2017/02/Optimism-and-Pessimism-with-Expected-
Utility-3.2016-st4ebu.pdf

[260] Wasfi A. Al-Rawabdeh and Doraid Dalalah: Predictive decision making under risk and uncertainty: A support vector machines model,
RAIRO-Oper. Res. 51 (2017) 639–667.

[261] Keun TaeCho: Multicriteria decision methods: An attempt to evaluate
and unify, http://www.sciencedirect.com/science/article/pii/S0895717
703001225

[262] Darius Braziunas: POMDP solution methods, A survey, 2003,
http://www.cs.toronto.edu/~darius/papers/POMDP_survey.pdf

[263] Pengfei Zhu, Xin Li, Pascal Poupart: On Improving Deep Reinforce-
ment Learning for POMDPs, https://arxiv.org/pdf/1704.07978.pdf

[264] George E. Monahan: State of the Art—A Survey of Partially Observable Markov Decision Processes: Theory, Models, and Algorithms,
1982, https://doi.org/10.1287/mnsc.28.1.1

[265] Leonore Winterer, Sebastian Junges, Ralf Wimmer, Nils Jansen,
Ufuk Topcu, Joost-Pieter Katoen, Bernd Becker: Motion Planning
under Partial Observability using Game-Based Abstraction, http://
www.cs.ru.nl/personal/nilsjansen/files/publications/winterer-et-al-cdc-
2017.pdf

[266] Roel Dobbe, David Fridovich-Keil, Claire Tomlin: Fully Decentralized Policies for Multi-Agent Systems: An Information Theoretic
Approach, August 1, 2017, https://arxiv.org/pdf/1707.06334.pdf

[267] Kai Xu, Kaiming Xiao, Quanjun Yin, Yabing Zha, Cheng Zhu:
Bridging the Gap between Observation and Decision Making:
Goal Recognition and Flexible Resource Allocation in Dynamic
Network Interdiction, Proceedings of the Twenty-Sixth Interna-
tional Joint Conference on Artificial Intelligence (IJCAI-17), 4477,
http://static.ijcai.org/proceedings-2017/0625.pdf

[268] Robert Platt, Leslie Kaelbling, Tomas Lozano-Perez, Russ Tedrake: Non-Gaussian Belief Space Planning: Correctness and Complexity, http://lis.csail.mit.edu/pubs/tlp/ICRA12_1775_FI.pdf

[269] Zi Wang, Stefanie Jegelka, Leslie Kaelbling, Tomas Lozano-Perez: Focused Model-Learning and Planning for Non-Gaussian Continuous State-Action Systems, 2017, http://lis.csail.mit.edu/pubs/wang-icra17.pdf

[270] E. Pashajavid, M.A. Golkar: Non-Gaussian multivariate modeling of plug-in electric vehicles load demand, http://dx.doi.org/10.1016/j.ijepes.2014.03.021

[271] Finn Müller-Hansen et al.: How to represent human behaviour and decision making in Earth system models? A guide to techniques and approaches, *Earth Syst. Dynam. Discuss.*, doi:10.5194/esd-2017-18, 2017

[272] Vyacheslav I. Yukalov and Didier Sornette: Quantum Probabilities as Behavioral Probabilities, Entropy 2017, 19, 112; doi:10.3390/e19030112

[273] Frederic Moisan, Robert ten Brinckehttp: Not all Prisoner's Dilemma Games are Equal: Incentives, Social Preferences, and Cooperation, www.hss.cmu.edu/departments/sds/ddmlab/papers/MoisantenBrincke etal.pdf

[274] V. I. Yukalov, D. Sornette: Quantum probability and quantum decision-making, November 2015, DOI: 10.1098/rsta.2015.0100

[275] Tobias U. Hauser, Micah Allen, Geraint Rees, Raymond J.: Metacognitive impairments extend perceptual decision making weaknesses in compulsivity, http://dx.doi.org/10.1101/098277, 2017

[276] P. Kravets: NEUROAGENT MODEL OF DECISION-MAKING, UDC 004.032.26; 004.852; 004.94, http://science.lpnu.ua/sites/default/files/journal-paper/2017/may/2387/kravetspetroenglish.pdf

[277] Nick Yeung, Christopher Summerfield: Metacognition in human decision-making: confidence and error monitoring, 9 April 2012, DOI: 10.1098/rstb.2011.0416, http://rstb.royalsocietypublishing.org/content/367/1594/1310.short

[278] Luc Boerboom & Valentina Ferrett: Actor-Network-Theory perspective on a forestry decision support system design, Scandinavian Journal of Forest Research, Volume 29, 2014, http://www.tandfonline.com/doi/abs/10.1080/02827581.2014.946960

[279] Sofia Maria Dima, Christos Antonopoulos, and Stavros Koubias: Resource Aware Sensor-to-Actor Allocation Framework for WSANs

Based on Voronoi Cells Theory Journal of Sensors Volume 2017, Article ID 4523945, 16 pages, https://doi.org/10.1155/2017/4523945

[280] Levent Yilmaz: VERIFICATION AND VALIDATION OF ETHICAL DECISION-MAKING IN AUTONOMOUS SYSTEMS, http://scs. org/wp-content/uploads/2017/06/1_Final_Manuscript-3.pdf

[281] GLT: https://football-technology.fifa.com/media/1010/goal-line_tech nology_testing_manual_2012.pdf

[282] Paolo Spagnolo, Marco Leo, Pier Luigi Mazzeo, Massimiliano Nitti, Ettore Stella, Arcangelo: Non-Invasive Soccer Goal Line Technology: A Real Case Study, 2013 IEEE Conference on Computer Vision and Pattern Recognition Workshops, June 2013, DOI: 10.1109/CVPRW.2013.147

[283] Abbas Mardani, Ahmad Jusoh, Khalil MD Nor, Zainab Khalifah, Norhayati Zakwan & Alireza Valipour: Multiple criteria decision-making techniques and their applications – a review of the literature from 2000 to 2014, Economic Research-Ekonomska Istraživanja, 2015 Vol. 28, No. 1, 516–571, http://dx.doi.org/10.1080/1331677X.2015.10 75139

[284] Markus Holzer, Martin Kutrib: Descriptional and computational complexity of finite automata—A survey, Information and Computation 209 (2011) 456–470, http://ac.els-cdn.com/S0890540110001999/1-s 2.0-S0890540110001999-main.pdf?_tid=71a72df4-89f7-11e7-92c1-0 0000aacb360&acdnat=1503708331_2102223d2dc72aaff42a4ff2ff7f3 fe2

[285] Yo-Sub Hana, Arto Salomaab, and Kai Salomaac: Ambiguity, Nondeterminism and State Complexity of Finite Automata, Acta Cybernetica 23 (2017) 141–157, http://www.inf.u-szeged.hu/actacybernetica/edb/ vol23n1/pdf/actacyb_23_1_2017_9.pdf DOI: 10.14232/actacyb.23.1. 2017.9

[286] E.P. de Vink: Introduction to Automata and Process Theory, 2015, http://www.win.tue.nl/~hzantema/ap.pdf

[287] Lockheed Martin F-35, https://en.wikipedia.org/wiki/Lockheed_ Martin_F-35_Lightning_II

[288] Lawrence Chung: Non-Functional Requirements, https://www.utdallas .edu/~chung/SYSM6309/NFR-18-4-on-1.pdf

[289] Peter Eeles: Non-Functional Requirements, IBM UK, http://www. architecting.co.uk/presentations/NFRs.pdf

[290] Martin Glinz: On Non-Functional Requirements, https://pdfs.semantic scholar.org/3d87/83cb02ce71f1ff0d78a805f8374925885a55.pdf

[291] PLATFORM4CPS, EC funded project on CPS, https://www.platforms 4cps.eu/fileadmin/user_upload/Platforms4CPS_Newsletter_No._1.pdf

[292] Sara Bouchenak, Francesco Brancati, Andrea Ceccarelli, Sorin Iacob, Nicolas Marchand, et al. Managing Dynamicity in SoS. Cyber-Physical Systems of Systems: Foundations – A Conceptual Model and Some Derivations: The AMADEOS Legacy, 13, pp. 926–206, 2016, 978-3-319-47590-5

[293] Theodoridis, S.: Machine Learning: A Bayesian Optimisation Perspective. Elsevier, London (2015), http://www.sciencedirect.com/science/book/9780128015223

[294] Vegard Engen, J. Brian Pickering and Paul Walland: Machine Agency in Human-Machine Networks; Impacts and Trust Implications, Pre-print; 18th International Conference on Human-Computer Interaction International, Toronto, Canada, 17–22 July 2016 [HUMANE Project, EC], https://pdfs.semanticscholar.org/2dcc/c1bdbe8b28af72dece914 e0eb27bcce2e96b.pdf

[295] A ROADMAP FOR FUTURE HUMAN-MACHINE NETWORKS FOR CITIZEN PARTICIPATION, HUMANE Project, EC, https:// humane2020.eu/blog/

[296] Rozita Jamili Oskouei, Hamidreza Naghizadeh, Zahra Samadyar: Intelligent Agents: A Comprehensive Survey, International Journal of Electronics Communication and Computer Engineering Volume 5, Issue 4, ISSN (Online): 2249–071X, ISSN (Print): 2278–4209.

[297] Hannes Leitgeb: Introduction: Belief vs Degrees of Belief, 2014, https://www.tilburguniversity.edu/upload/8a7aa405-47ff-4c23-8f98-b25430631792_Presentation%20Leitgeb%20intro.pdf

2

On Designing Decision-Making Mechanisms for Cyber-Physical Systems

Kostas Siozios[1], Dimitrios Soudris[2] and Elias Kosmatopoulos[3]

[1]Department of Physics, Aristotle University of Thessaloniki, Greece
[2]School of ECE, National Technical University of Athens, Greece
[3]Department of ECE, Democritus University of Thrace, Greece

Abstract

A Cyber-Physical System (CPS) is considered as one of the hottest computer applications today, while a proper design of such a system preassumes that a number of challenges have to be sufficiently addressed. A CPS is composed by a tight integration of cyber and physical objects, where the term cyber objects refers to any computing hardware/software resources that can achieve computation, communication, and control functions in a discrete, logical, and switched environment. Similarly, the physical entities refer to any natural or human-made systems that are governed by the laws of physics and operate in continuous time. In order to address these challenges, three complementary technologies, namely sensing, computing, and communication have to be proper combined. Critical role to any CPS is the decision-making mechanism, which controls the individual entities/services in order all of them to be orchestrated and operate as a unique system. For this purpose, both the cyber and physical aspects of a CPS have to be appropriately designed, implemented, and customized in order to maximize the potential gains from these platforms.

2.1 Introduction

Over the years, a number of paradigms to describe the task of decision-making process have been proposed. It consists of four phases, namely intelligence, design, choice, and implementation, as it is shown in Figure 2.1.

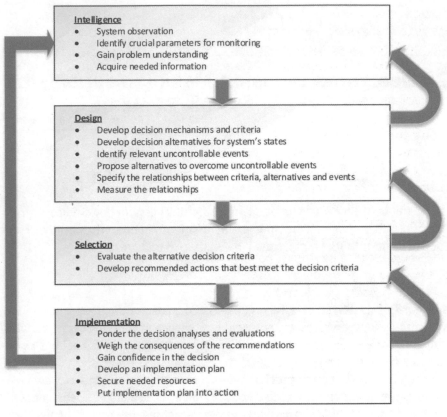

Figure 2.1   The process of decision making.

More precisely, in the intelligence phase, a decision maker observes the physical system, and establishes an understanding of the problem domain and the
associated opportunities. The necessary information pertaining to all aspects
of the problem under scrutiny is also collected. In the design phase, the
alternative decision criteria are developed. For this purpose, a specific model,
with the relevant uncontrollable events identified is employed. In order to
succeed this task, the relationships between the decisions, alternatives, and
events have to be clearly specified and measured. The quality of this task
is upmost important, as it usually dominates the performance of the overall
decision-making mechanism. This enables the decision alternatives to be
evaluated logically in the next phase, i.e., the choice phase. Besides, actions
that best meet the decision criteria are formulated. In the implementation
phase, the decision maker needs to re-consider the decision analyses and

evaluations, as well as to weigh the consequences of the recommendations. An implementation plan is then developed, with the necessary resources secured. Finally, the implementation plan is put into action. It is also important to notice that the decision-making process is a continuous task within a feedback loop. This means that the decision-making mechanism has to constantly re-consider and re-evaluate the reality and changes in the problem domain. On other words, new information is obtained by re-visiting one or more, if not all (depending on inherent constraints and limitations posed by the target application domain), of the four previously mentioned phases. The feedback process allows alterations and improvements on previous decisions to be accomplished, so as to meet the current needs and demands of the problem domain.

Recently, the convergence of emerging embedded computing, information technology, and distributed control became a key enabler for future technologies. Among others, a new generation of systems with integrated computational and physical capabilities that can interact with humans through many new modalities have been introduced. Furthermore, it is expected that computing and communication capabilities will soon be embedded in all types of objects and structures in the physical environment [1]. Applications with enormous societal impact and economic benefit will be created by harnessing these capabilities across both space and time domains. Such systems that bridge the cyber world of computing and communications with the physical world are referred to as CPSs. Specifically, a CPS is a collection of task-oriented or dedicated systems that pool their resources and capabilities together to create a new, more complex system, which offers more functionality and performance than simply the sum of the constituent sub-systems. Among others, these new design paradigms have the ability to interact with, and expand the capabilities of, the physical world through monitoring, computation, communication, coordination, and decision-making mechanisms, as it is depicted at Figure 2.2. Such an emerging multidisciplinary frontier will enable revolutionary changes in the way humans live, while it is also expected to be a key enabler for future technology developments.

The integration of physical processes and computing is not new. Embedded systems have been in place for a long time and these systems often combine physical processes (e.g., through digital/analog sensors) with computing. However, the core differentiator between a CPS and either a typical control system, or an embedded system, is the communication feature among system's components, which adds reconfigurability and scalability, allowing instrumenting the physical world with pervasive networks of sensor-rich

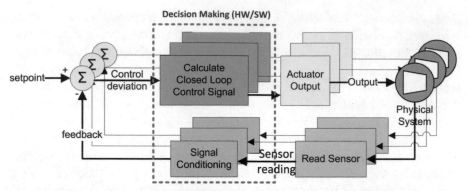

Figure 2.2 The block diagram of a typical CPS.

embedded computation [2]. The goal of CPS architecture is to get maximum value out of a large system by understanding of how each of the smaller (sub-) systems work, interface and interact. This trend is also supported by the continuation of Moore's law, which imposes that the cost of a single embedded computer equipped with sensing, processing and communication capabilities drops towards zero [3]. Thus, it is economically feasible to densely deploy networks with very large quantities of sensor readings from the physical world, compute quantities and take decisions out of them. Such a very dense network offer a better resolution of the physical world and therefore a better capability of detecting the occurrence of an event; this is of paramount importance for a number of foreseeable applications.

The availability of wide-range architectural components for supporting the sensing, processing and communication capabilities make the implementation of CPSs a viable solution. However, the efficiency of final products mainly relies on the flexible, yet efficient, algorithms for supporting the system's functionality, which mainly provides some form of orchestrating systems components. Previous studies indicate that decision making (i.e., control mechanisms) is though as the largest obstacle towards the design of an efficient dynamical system [4, 5]. Specifically, since the orchestrator is a system's component (or a set of components) that manages, commands, directs or regulates the behavior of other entities, it highly affects almost the majority of performance metrics. The ancient Greek philosopher Aristotle, already started to think about the decision making. Following his statement "... *if every instrument could accomplish its own work, obeying or anticipating the will of others ... chief workmen would not want servants, nor masters slaves* ...", he tried to automatize the behavior of others (e.g., people, devices) using a set of clearly defined criteria. This is also the core idea behind the development of CPSs — to force them so that they work in a manner they are designed. If so, quality assurance for this development gains the priority too.

The design and implementation of an efficient orchestrator of CPS is a challenging task, as real-time constraints have usually to be met. Moreover, the inherent heterogeneity found in modular components of the target system further complicates this problem. To make matters worse, existing approaches for CPS usually assume components with increased processing power, which is not always the case especially at the embedded domain, where emphasis is paid mainly in ultra low-power solutions. Finally, as the physical world is not entirely predictable, it is not expected the CPS to be operating in a controlled environment; thus, it must be robust enough to unexpected conditions and adaptable to subsystem failures.

Due to its importance, control theory was developed as an interdisciplinary branch of engineering and mathematics that deals with the behavior of dynamic systems. At this theory, the desired output of a system is called the *reference*. In case one (or more) output variables of a system need to follow a certain reference over time, there is a controller that manipulates the inputs to obtain the desired effect on the system's output. Whenever the system imposes dynamic behavior, then usually it is controlled with a feedback loop [6]. This feedback is a process whereby some proportion of the output signal of a system is passed (feedback) to the input. A further enhancement to this control is achieved with the closed-loop controller, where the feedback controls either states or outputs of a dynamic system [4]. The advantages of closed-loop controllers, as compared to the corresponding open-loop controllers are summarized as follows: they have guaranteed performance, they reject system's disturbance (e.g., unmeasured friction in a motor), they exhibit capability to stabilize unstable processes, they have reduced sensitivity to parameter variations, and they exhibit improved performance for reference tracking.

Apart from the efficiency of control mechanisms discussed so far, the majority of them exhibit increased computational and/or storage complexity, which in turn makes their implementation as part of an embedded system a challenging issue [5]. The importance of this problem was also highlighted as a big challenge for upcoming large-scale CPS platforms [1]. Note that the absence of developing lightweight solutions (able to be executed even onto low-power embedded platforms) for supporting the task of decision making for CPS is not due to neglect, but rather due to its difficulty. Also, as we have already mentioned, the problem of orchestrating heterogeneous system's components becomes far more challenging in case the decision-making is performed under run-time (or real-time) constraints. In such a case, usually a compromise between the desired accuracy and the processing overhead is applied.

Existing approaches for supporting the systems decision making rely mainly on *ad hoc* methods: After all the components have been designed and manufactured, the control mechanisms aim at making the system to work somehow. However, as the complexity of engineered systems continues to increase, the lack of a systematic theory for orchestrating the functionality of entire CPS platform introduces additional limitations. Toward this direction, throughout this manuscript we propose a framework for supporting rapid prototyping of decision-making mechanisms targeting to CPS. The proposed framework employees a simulation layer in order to enable part(s) of the physical components to be replaced with their equivalent software models without distributing the overall system's functionality.

Among others, this framework takes into consideration the interactions between systems control and sensors/actuators in order to ensure both system-level properties, as well as the system-level behavioral simulation with incomplete (or limited) information. Additionally, we discuss the associated open-source toolbox that automates the tasks of design space exploration (i.e., evaluate the candidate solutions from the search space) and instantiation of the selected decision-making approach according to system's specifications. For demonstration purposes, the proposed design framework is employed to support the design of decision-making mechanisms targeting to smart thermostats. During the development phase, data from both simulated and real-world data from weather stations were considered. By enabling the control infrastructure to support the dynamic configuration of thermostats our solution achieves to maximize the thermal comfort with the minimum energy dissipation, as compared to existing static approaches. Moreover, it is almost negligible requirements for processing power and storage makes the designed decision-making approach suitable for execution on embedded platforms, as compared to the majority of existing control techniques that aims solely on accuracy (ignoring issues related to physical implementation).

2.2 Related Work for Designing CPS Platforms

The complexity of CPS platforms imposes the requirement for using auto-mated techniques during the specification, implementation, and testing phases of such systems. For this purpose, all the necessary high-confidence mechanisms that are able to interact with humans and the physical world in dynamic environments and under unforeseen conditions have to be employed. Furthermore, the number of components found at these systems, in conjunction to their diverse functionalities, impose that key role at design phase of CPS is given to an efficient system's modeling and co-simulation phase. Thus, careful analysis of the well-established techniques at the domain of rapid prototyping for CPSs is necessary in order to identify open issues and propose the corresponding solutions.

Both academia and industry has already identified the necessity for addressing this goal. Inline to this objective various model of computation and communication have to be considered, as both of them dominate the performance of CPS platforms. Apart from these models, simulation tools are upmost important as they enable accurate exploration of the design space, as well as the overall system's validation in terms of functional and timing aspects. Note that simulators are also widely used for verifying the proper functionality of individual components (either analog, or digital). As a consequence of this effort, a number of different simulation approaches have been proposed that trade-off accuracy with complexity. Thanks to these solutions, the hardware components of a CPS system can be simulated at different levels of abstraction (e.g., electrical circuits, logic gates, or register-transfer level). Regarding the functional validation of the system's behaviour, it is possible to be performed with a formal specification language, such as the Statecharts or the Specification and Description Language (SDL). However, at this layer of abstraction it is not feasible to consider either the system's timing behavior, or the interactions between the applications device drivers with the underline hardware.

In order to overcome the limitation between the system's specification and its implementation phase (which will enable the efficient co-design and validation of large-scale systems in the real environment), it is common to employ rapid prototyping techniques. Beside this, the increasing software content-on-time and on-budget requires starting the software development process as early as possible. Today, designers use two relatively independent methods for supporting rapid prototyping: TLM (Transaction-Level Model)-based virtual prototyping and FPGA-based prototyping. Virtual prototypes are fast, fully functional models of SoCs under development (described at SystemC or Transaction Level Model) that execute unmodified production code. Because of unparalleled software debug and analysis efficiency, virtual prototyping accelerates pre-RTL (Register Transfer Level) embedded software development, hardware/software integration, and system validation. On the other hand, FPGA-based prototyping accelerates the creation of an ASIC prototype with high-speed hardware prototyping systems including a software flow for the conversion of ASIC RTL into one or more FPGA devices. This type of prototypes provides cycle-accurate, near real-time performance, and real world interface connectivity prior to availability of first platforms silicon. Although both virtual and FPGA-based prototyping approaches have unique benefits to the development team, the mixing of SystemC and RTL model abstractions is a challenging task. Thus, in order to accelerate time-to-market, more advanced ASIC prototyping techniques, such as a hybrid prototype are being adopted to enable designers to start multi-core SoC prototyping earlier.

Industry has already identified this trend, as there are various environments for supporting scalable prototyping systems, such as the HAPS [7], Hybrid Prototyping [8], Certify [9], Symplify Premier [10] and Protium [11]. The majority of these solutions rely either on a single FPGA, or an array of FPGAs, in order to accelerate the rapid prototyping phase. Even though these frameworks lead to high-quality solutions, they focus mainly to digital platforms (e.g., embedded systems). However, as the CPS involves also interaction to physical world, more advanced frameworks are absolutely necessary to tackle the sufficient design of these systems. Among others such frameworks have to provide sufficient signal/data transactions between physical and cyber environments. Furthermore, apart from the previously mentioned methodologies, software tools have also to be developed in order to tackle the increased complexity of CPS. Since it is very difficult (and maybe inefficient) to develop an environment for supporting tasks related to rapid prototyping of CPS from scratch, up to now effort was mostly paid to develop the appropriate interfaces among existing techniques/tools that address parts of the entire problem.

A methodology that integrates both discrete and continuous time models at different layers of abstraction (SystemC and Simulink) are discussed in [12]. A similar framework is presented in [5], where the co-simulation framework integrates VDM++ and 20-sim. Although these approaches exhibit remarkable results, they are not directly applicable to the design of CPS as they do not provide sufficient interaction between cyber and physical world. In Synopsys [13], a methodology of virtual prototyping is proposed which combines SystemC, QEMU, and Open Dynamics Engine to achieve a holistic design view. The TrueTime toolbox (developed in MATLAB/Simulink environment) is a relevant approach for enabling CPS simulation, which considers timing aspects introduced by computation and communication [14]. Unfortunately, this toolbox cannot integrate hardware models, while it does not support different abstraction levels. A co-simulation framework that deals with the joint design of software in C, hardware in HDL and mechanical components in MATLAB is discussed in Verhoef et al. [15].

The previous analysis indicates the importance for tools that handle efficiently the increased complexity imposed by the rapid prototyping of CPS. For this purpose, methodologies, as well as the associated tool flows that perform flexible (in terms of accuracy and computational complexity) simulation, debugging and verification of the entire system are absolutely necessary. This problem becomes more challenging by taking into consideration that existing flows are built on the fundamental premise that models are freely interchangeable amongst vendors and have interoperability amongst them. In other words, this claim imposes that models can be written, or obtained from other vendors, while it is known *a priori* that they will be accepted by any vendor tool for performing different steps of system prototyping (e.g., architecture's analysis, modeling, simulation, etc.). Even though this concept seems straightforward and promising, it has proven completely elusive, since the existing design solutions do not provide either model interoperability or independence between model and software tools. Also, the previously mentioned analysis indicates that there is a plenty of simulators and design environments that tackle partially the task of rapid prototyping of digital/analog systems at different layers of abstraction. However, none of them is directly applicable to the CPS domain, as they cannot support cross-domain concepts for architecture, communication and compatibility. For this purpose, novel approaches that can be classified at two complementary directions are absolutely necessary: (i) develop frameworks that cyberize the physical systems means to endow physical subsystems with cyber-like abstractions and interfaces (wrap software abstractions around physical subsystems); and (ii) develop frameworks that physicalize the cyber means to endow software and network components with abstractions and interfaces that represent their dynamics in time.

2.3 Conclusions

An overview of hardware/software techniques employed for the sufficiently design of decision-making mechanisms was introduced. Due to the complexity of these techniques, a number of software tools are employed.

References

[1] Hipeac. (2015). *Hipeac vision.* Available at: https://www.hipeac.net/assets/public/publications/vision/hipeac-vision-2015_Dq0boL8.pdf (accessed 30 April, 2017].

[2] Estrin, D., Culler, D., Pister, K., and Sukhatme G. (2002). Connecting the physical world with pervasive networks. *IEEE Pervasive Comput.* 1, 59–69.

[3] ITRS. (2013). *International Technology Roadmap for Semiconductors,* Washington, DC.

[4] Kilian, C. T. (1996). *Modern Control Technology: Components and Systems,* 1st edn. St. Paul, MN: West Publishing Co.

[5] Verhoef, M., Visser, P., Hooman, J., and Broenink, J. (2007). *Co-simulation of Distributed Embedded Real-Time Control Systems* (Berlin: Springer), 639–658.

[6] Spencer, R. R. and Ghausi, M. S. (2003). *Introduction to Electronic Circuit Design.* Upper River Sadle, NJ: Prentice Hall.

[7] Synopsys. (2017). Haps FPGA-based prototyping solutions. https://www.synopsys.com/Prototyping/FPGABasedPrototyping/Pages/HAPS.aspx (accessed 30 April 2017).

[8] Synopsys. (2017). Synopsys hybrid prototyping. Available at: https://www.synopsys.com/Prototyping/FPGABasedPrototyping/Pages/hybrid-prototyping.aspx (accessed 30 April 2017).

[9] Synopsys. (2017). Implement high-performance fpga-based asic prototypes with synopsys certify. Available at: https://www.synopsys.com/Prototyping/FPGABasedPrototyping/Pages/certify.aspx (accessed 30 April 2017).

[10] Synopsys. (2017). Fast implementation of advanced fpga designs and fpga-based prototypes with synopsys synplify premier. Available at: https://www.synopsys.com/Tools/Implementation/FPGAImplementation/FPGASynthesis/Pages/SynplifyPremier.aspx (accessed 30 April 2017).

[11] Cadence. (2017). *Protium rapid prototyping platform.* Available at: http://www.cadence.com/products/sd/protium_rapid_prototyping/pages/default.aspx (accessed 30 April 2017).

[12] Gheorghe, L., Bouchhima, F., Nicolescu, G., and Boucheneb, H. (2006). "Formal definitions of simulation interfaces in a continuous/discrete co-simulation tool," in *Seventeenth IEEE International Workshop on Rapid System Prototyping (RSP'06)* (New York, NY: IEEE), 186–192.

[13] Mueller, W., Becker, M., Elfeky, A., and DiPasquale A. (2012). "Virtual prototyping of cyber-physical systems," in *17th Asia and South Pacific Design Automation Conference* (New York, NY: IEEE), 219–226.

[14] Cervin, A., Henriksson, D., Lincoln, B., Eker, J., and Arzen, K. E. (2003). How does control timing affect performance? analysis and simulation of timing using jitterbug and truetime. *IEEE Control Syst.* 23, 16–30.

[15] Le Marrec, P., Valderrama, C. A., Hessel, F., Jerraya, Attia, M., and Cayrol, O. (1998). "Hardware, software and mechanical cosimulation for automotive applications," in *Proceedings. Ninth International Workshop on Rapid System Prototyping (Cat. No.98TB100237)* (New York, NY: IEEE), 202–206.

3

Design Space Exploration Methodology Based on Decision Trees for Cyber-Physical Systems

Lazaros Papadopoulos and Dimitrios Soudris

School of Electrical and Computer Engineering, National Technical University of Athens, Athens, Greece

Abstract

As cyber-physical systems become increasingly complex, effective design space exploration becomes more and more challenging. Exploration parameters vary a lot between different architectures, while the design spaces are often large and complex. This work presents a systematic way for defining design spaces and it is based on a decision tree representation. The proposed methodology assists the effective exploration and the development of tools that support it. It is demonstrated in two different use cases in the context of cyber-physical systems.

Introduction

Design space exploration (DSE) is a widely used approach that is based on the exploration of a set of design options, the evaluation of each one and the selection of the most efficient. It is often used for optimizing embedded systems by achieving trade-offs between design metrics.

Many recent works leverage DSE for effective HW/SW co-design [1, 2] and often target performance vs. energy trade-offs at various abstraction levels. For example, some works focus on parameter tuning at architectural level [3, 4], while others focus on system-level design [2, 5]. For instance, DSE of heterogeneous embedded architectures is presented in Palermo et al. [5].

111

The methodology is supported by a tool chain for rapid system design that focuses on high performance and low energy consumption. Other works focus on source-to-source transformations for customization of dynamic data structures [6] and of dynamic memory management optimization [7]. Finally, a survey on DSE in embedded systems is presented in [8].

For DSE to be effective, it is necessary the set of design options to be represented in a consistent and systematic way. In this work, we present a DSE methodology for cyber-physical systems, which is also applicable to any other embedded platform. The design options of a platform-independent design space are presented as a set of decision trees. A set of constraints "convert" the design space from platform-independent to architectural-specific, by pruning the non-applicable options. Finally, a set of interdependency rules are used to instantiate the implementations (i.e., solutions) that should be forwarded to the exploration process.

The methodology has several advantages: first, it provides a systematic way to represent the design space. The set of constrains prune the design space, thus decreasing its size and avoiding the evaluation of solutions that are practically non-applicable or provide poor performance results. Also, such a systematic representation assists the development of tools that automate the whole DSE process, as much as possible. Finally, the methodology can be applied in several problems in the area of embedded systems, especially in cases in which the design space is complex.

The components and the flow of the methodology are presented in Section 3.1. The design options, the constraints and the concept of interdependencies are analysed in detail. The implementation of the methodology in two use cases in which DSE is required is presented in Section 3.2. First, it is demonstrated in the design space of the concurrent data structures and then in the implementation of the multiway streaming aggregation in embedded architectures. Finally, in Section 3.3, we draw the conclusions.

3.1 Methodology

The proposed methodology consists of three components: The design options, the constraints and the interdependencies. They will be presented in detail in this section, along with the methodology flow.

3.1.1 Design Options

The design options can be represented in a systematic way through a set of decision trees. Each decision tree has a number of leaves and each leaf takes a specific value. Each unique combination of leaves corresponds to a single solution that should be evaluated.

The design options describe parameters, characteristics, or features of the solutions and comprise the design space. In other words, we leverage the concept of decision trees in order to organize the design space in a set of simple design decisions. The value that each leaf of the decision trees takes can be continuous or discrete (e.g., yes/no), depending on the context in which the methodology is applied. Figure 3.1 shows a set of design options represented as decision trees. Each design option (O_1, O_2, ... O_N) has a number of leaves that each one takes a different value (O_{1a}, O_{1b}, etc.).

3.1.2 Constraints

Not every design option is applicable in any context. Some options may be non-applicable in specific contexts, being irrelevant or meaningless. Therefore, each decision tree, apart from the values related to the feature or the characteristic that it represents (which are the tree leaves), has one more set

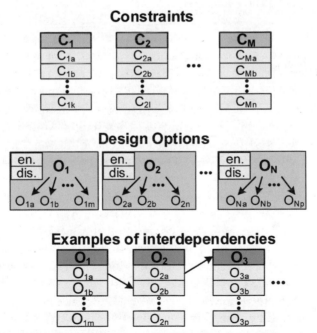

Figure 3.1 Components of the methodology: constraints, design options and interdependencies.

of values that are mutually exclusive: enabled and disabled, shown as *en.* and *dis.* in each decision tree of Figure 3.1. To determine the availability of the options in different situations, we use a set of constraints that affect the applicability of the design options in each specific context. Similarly, with the design options, the constraints can take different values. Each single combination of constraint values represents a different context in which the system operates. In Figure 3.1, M constraints are shown, each one having a set of possible values that it can take. Therefore, we assigned each single value of each constraint with the enabled or the disabled value of each option. By applying a set of constraint values that represent a specific context, the non-applicable options are disabled (i.e. removed), leaving only the ones that are relevant with the context in which the system is executed. Thus, the total number of identified solutions can be significantly reduced, since the fact that the non-applicable options are pruned, results in the instantiation of the valid and meaningful solutions only.

Finally, it is important to state that the type of constraints, their values and the rules by which each constraint is assigned to the *en.* or *dis.* value of each decision tree depend on the context in which the methodology is applied, entirely. The user that applies the proposed methodology on a specific DSE problem sets the constraints and the associated rules.

3.1.3 Interdependencies

The design options are not independent, in the sense that a specific value of a design option can make another design option (or multiple options) or leaf/leaves non-applicable or meaningless. Therefore, it is important to define a set of rules which determine how each decision tree and/or leaf affects the applicability of all the others. These rules define the *interdependencies* between the design options. In Figure 3.1, an interdependency is represented with an arrow pointing from decision tree leaf O_{1a} to decision tree leaf O_{2b}. This means that if the leaf O_{1a} is selected during the instantiation of a solution, then leaf O_{2b} cannot be used during the instantiation of the current solution. Another interdependency rule presented as an example in Figure 3.1, shows that if during the instantiation of a solution leaf O_{2a} is selected, then the whole decision tree O_3 is disabled.

It is important to differentiate between the constraints and the inter-dependency rules. The constraints are used to "convert" the design space from generic and platform-independent, to platform-specific. By removing the non-applicable design options and the ones that are expected to provide very poor performance results, the size of the design space is reduced. Also, the constraints are applied at decision tree level and not at leaf level. On the other hand, the role of interdependencies is different: they describe the set of rules for the instantiation of the solutions from a set of decision trees that comprise an already platform-specific design space.

3.1.4 Methodology Flow

The implementation of the methodology in a specific context is a three-step procedure and it is shown in Figure 3.2. The first step is the representation of the design options in a decision tree form. The different values that each design option can take are the leaves of each tree. The second step is the identification of the constraints and the corresponding decision trees that are enabled or disabled for each combination of constraint values. Finally, the last step is the definition of the interdependency rules, which determine the instantiation of each produced solution.

After the implementation of the above three steps in a specific context, the methodology can be applied in the same context for different constraint values. This is illustrated in Figure 3.3. Different set of constraints can be applied as an input to the methodology. Then, each decision tree is enabled or disabled, according to the predefined rules. The interdependency rules will be used to instantiate a set of solutions. These solutions will be explored and evaluated in order to identify the optimal one. Therefore, when the constraints

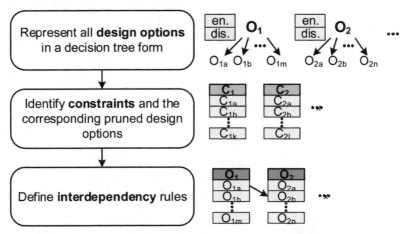

Figure 3.2 Instantiation of the methodology in a specific context.

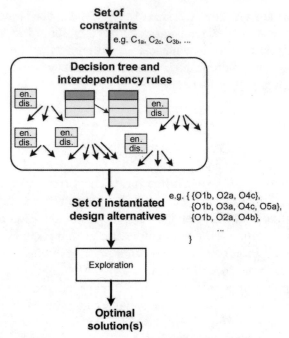

Figure 3.3 Implementation of the methodology.

change, there is no need to redefine the design space. Instead, the predefined rules adapt it, by enabling and disabling the corresponding decision trees.

An important feature of the proposed methodology is that it can be extended when the design space is expanded, in a systematic way. In order to add a new design option in the design space, a three-step procedure should be followed. The first step is the extension of the design space that it can take place either by the addition of new leaves to existing decision trees, or by adding a new decision tree. The second step, is the update of the design constraints. It should be determined whether or not the new design option is disabled for specific constraint values. It is important to state that a constraint cannot prune one or more decision tree leaves. It can only prune one or more complete decision trees. This is important for ensuring the coherence of the methodology. The third step is to identify the interdependences between the new and the existing design options and update the interdependency rules.

The overhead of the methodology mostly depends on the number of design options in a specific context (i.e., on the number of parameters in a DSE problem). Apparently, when the number of decision trees increases, the number of the provided solutions increases as well and the exploration time is high. However, the pruning of the design space that takes place during the implementation of constraints and the interdependency rules reduces the number of solution and the exploration time, drastically.

To summarize, the goal of the methodology is to provide a systematic DSE methodology that can be applied in various contexts, especially when the design space is large and complex.

- It is a systematic way to represent the design space and to automate steps of the exploration process.
- The methodology provides the automatic pruning of the design options that are non-applicable in a specific context (or are expected to provide poor results) through a set of predefined rules. Thus, the design space is reduced and the exploration time decreases.
- The methodology provides extensibility features. Thus, new design options can be added to the design space in a systematic way.
- Finally, it can be applied to several DSE problems, in various contexts with large design space.

3.2 Demonstration of the Methodology

In this section, we demonstrate the implementation of the methodology in two different scenarios from the embedded systems domain, in which exploration is required due to the large design spaces. The first is the selection of effective concurrent data structures in embedded applications and the second the implementation of the multiway streaming aggregation in embedded architectures. In both implementations we present results on platforms with different characteristics and, therefore, different constraints.

3.2.1 DSE on Concurrent Data Structures

Several works in the literature show that the effective selection of data structures in embedded applications can increase the performance, the memory utilization and the energy efficiency of the whole system [6, 9]. However, the design space of the data structures is complex, therefore, the effective selection in not a trivial task. The implementation of the methodology in this context can provide a semi-automatic solution to the selection of data structures for embedded applications. In this work, we specifically focus on concurrent data structures.

Figure 3.4 shows the design options in the context of concurrent data structures for applications executed in embedded systems. The design options, presented as decision trees, are grouped in five categories:

- *Data Structure Design Decisions* category refers to the design of a concurrent data structure, without taking into consideration the synchronization algorithm that handles concurrent accesses.
- *Pthread Lock Decisions* category consists of the simple and the readers/writer pthread locks options.
- *Test-and-Set Lock Decisions* category groups the customization options for TAS and TTAS locks. Back-off policy decision determines whether or not a thread will withdraw for an amount of time (which can increase linearly or exponentially) from spinning on an occupied lock to reduce bus congestion.
- *Locking Granularity Decisions* category refers to the granularity of locks. Lock-striping is a well-known locking technique proposed for concurrent hash tables, in which each lock protects a range of the data structure elements.
- *Lock-less Synchronization Decisions* groups the design options for lock-less data structures.

The design space of Figure 3.4 is platform and application independent. Not all design options are meaningful for any application, nor available in all platforms. Before the exploration, it is necessary to prune the design space, by eliminating meaningless and non-supported design options.

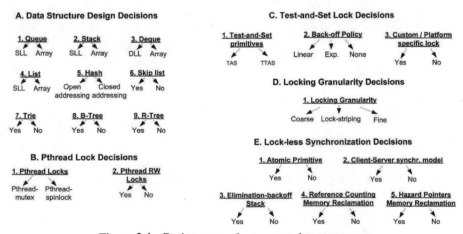

Figure 3.4 Design space of concurrent data structures.

We identified the application and the platform constraints that enable or disable specific categories or decision trees of the design space: The *access pattern* is an application constraint. The *number of threads* can be an application or a hardware constraint, while the *synchronization primitives support* is a hardware constraint. The constraints are the input of the methodology, provided by the user that results in the design space pruning at category or decision tree level (i.e., they do not prune specific leaves of design trees). This is important, in order to retain the coherence of the methodology and the tools that support it.

The **access pattern** is defined as the sequence that the data are accessed in a data structure by the application's algorithm. It refers to an abstraction level above the underlying data structure implementation. It affects the design decisions of category *A* of the design space and prunes meaningless abstract data structures. The access patterns and the corresponding enabled decision trees of Category *A* are shown in Table 3.1.

The **number of threads** constraint is related to the concurrency of the data structure and it can have two values: *one* and *many*. It determines if specific elements of the data structure are updated by multiple threads simultaneously. If the number is one, then categories *B*, *C*, *D*, and *E* are disabled, as shown in Table 3.2. In this case, the data structure is sequential and the exploration is limited to the decision trees of Category *A*.

The last constraint is the **synchronization primitives support**, which is hardware related and is presented in Table 3.3. The specifications of the platform enable decision trees from categories *B*, *C*, *D*, and *E*.

The interdependency rules are manually defined to instantiate coherent data structure implementations. As an example, in Figure 3.4, the selection of one decision tree of category *A* disables all other decision trees. The same applies to the decision trees of category *B*.

Table 3.1 Access pattern and corresponding enabled decision trees of the design space

Access Pattern	Enabled Decision Trees of Category 'A'
FIFO	A1
LIFO	A2
Deque	A3
Simple storage	A4
Key-value pairs storage	A5, A6, A8
Key-value pairs sorted storage	A6, A8
String storage	A7
Spacial access	A9

Table 3.2   Number of threads and corresponding disabled categories of the design space

Number of Threads	Disabled Categories
One	B, C, D, E
Many	–

Table 3.3   Synchronization primitives support and corresponding enabled Decision Trees of the design space

Synchronization Primitives Support	Enabled Decision Trees
Pthreads	B, D
Test-and-Set	C1, C2, D
Custom/Platf. Spec. locks	C3, D
Atomic Primitives with CAS	E1, E5
Atomic Primitives with DCAS	E1, E4
Message passing communication	E2

The methodology in the context of concurrent data structures consists of two steps and it is presented in Figure 3.5. The first step is the Concurrent Data Structure exploration and the second step is the Optimal Implementation

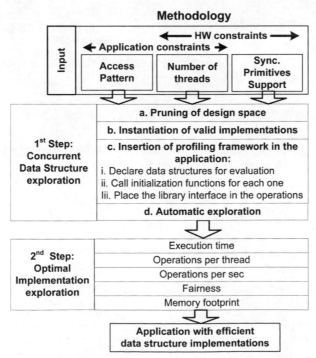

Figure 3.5   Instantiation of the methodology in the context of concurrent data structures.

exploration. The instantiation of the methodology in the context of concurrent data structures is supported by a library of data structures that it is developed to assist the exploration process.

To apply the first step of the methodology, the user provides the application and the hardware constraints information that prevents the evaluation of non-coherent and unsupported data structures. The pruned design space is provided to a script that handles the instantiation of valid data structure implementations, through a library of concurrent data structures that the toolchain that supports the exploration methodology provides. All remaining design choices are used to instantiate coherent data structures, according to the interdependences.

After the instantiation of the valid data structure implementations, developers insert the library interface to the application manually, by replacing the application's data structures under optimization with the ones of the library. Thus, all operations (e.g., insert, remove, etc.) pass through the library's data structures. Since data structures that correspond to the same access pattern (Table 3.1) have the same interface, this procedure needs to be done only once. Then, the exploration takes place: The application is executed for all different data structure implementations that were instantiated previously.

The second step of the methodology is the Optimal Implementation exploration. The profiling results for each concurrent data structure implementation that was evaluated are provided to developers. Taking into consideration the design constraints developers select the most efficient data structure implementation for the application under optimization.

Figure 3.6 shows an example of the results that are the outcome of the implementation of the methodology in the context of concurrent

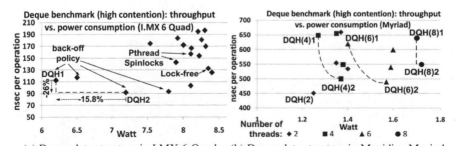

(a) Deque data structure in I.MX.6 Quad.   (b) Deque data structure in Movidius Myriad.

Figure 3.6 Demonstration of deque data structure on I.MX 6 Quad and on Movidius Myriad architectures.

data structures. It presents exploration results of performance vs. average power consumption for different implementations of the concurrent deque data structure in two platforms: in the ARM-based I.MX.6 Quad [10] (Figure 3.6(a)) and in the computer vision processor Myriad1 (Figure 3.6(b)) [11]. The methodology assists the development of a tool-flow that semi-automatically provides the exploration results in a convenient form, as shown in Figure 3.6. Developers can select the most effective deque implementation, according to the design constraints. More information about the instantiation of the proposed methodology can be found in [12].

3.2.2 DSE on Multiway Streaming Aggregation

Streaming aggregation is a widely used operator in the context of stream processing [13]. The effective implementation of the aggregation of multiple incoming streams in real time in embedded architectures is challenging, mainly for two reasons: First, the design space is large and complex, since parameters such as buffer sizes, DMA sizes, data structure allocation, etc. should be explored. Second, the implementation should be tuned in order to minimize the latency as much as possible. By applying the methodology in the context of multiway streaming aggregation, the most efficient configuration can be selected.

The design space of the streaming aggregation implementation is presented as a set of decision trees, grouped into two categories (Figure 3.7):

- *Category A* consists of decision trees that refer to memory configuration and allocation. Cache configuration options (private cache for each core or shared cache for all cores) are depicted in decision tree *A4*. *A5* is related with the dynamic memory allocation that can be based on freelists or in *malloc/free* system calls.
- In *category B* are assigned decision trees related to data movement and means by which accesses to shared resources are synchronized. The first three decision trees refer to different ways that data can be copied from global to local memories, or from one local memory to another (depending on the embedded system's memory hierarchy). Decision trees *B4* and *B5* are about synchronization between Processing Elements (PEs), when accessing shared buffers. At low level, synchronization can be accomplished by spinning on shared variables (i.e. busy waiting) or by using other platform specific solutions. In platforms that run OS and support POSIX threads developers can utilized semaphores or monitors.

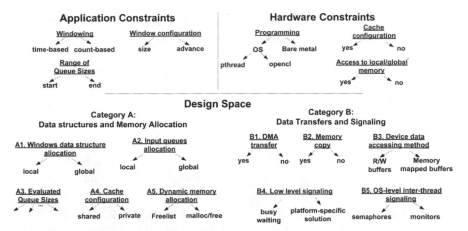

Figure 3.7 Design space of multiway streaming aggregation implementation in embedded devices.

Apparently, not all design options are applicable in any context. Figure 3.7 shows the application and the hardware constraints that affect which decision trees or leaves are applicable in each specific context. The constraints are used to prune the decision trees and leaves that yield implementations which do not adhere to developer's requirements or they are not supported by the embedded platform.

Table 3.4 summarizes the design options that are disabled, due to application and hardware constraints. As an example, if the embedded platform runs an OS, access to DMA and to low-level signaling mechanisms are most likely handled by the OS directly, so these design options are not exposed to developers. *Window configuration* constraint may force the allocation of the data structures in a global memory. All constraints are provided manually. Constraints that prune non-compatible design space options "convert" the

Table 3.4 Decision trees or leaves disabled for each application and hardware constraint

App./Hw Constraint	Decision Tree/ Leaf Disabled
Windowing(tuple-based)	$A2, A3, A5, B4$
Window configuration	may disable $A1$(local)
Programming(bare metal)	$B3$ and $B5$
Programming(pthread)	$B1, B3, B4$
Programming(OpenCL)	$B1, B2, B4, B5$
Cache config.(no)	$A4$
Access to local/global(no)	$A1, A2$

platform-independent design space into platform-dependent. Thus, they make the customization approach applicable in different contexts and in various embedded platforms.

After the pruning, valid customized streaming aggregation implementations are instantiated from the remaining decision tree leaves of the design space through the interdependency rules. In other words, the implementations that will finally be explored are the ones that are produced by combining the remaining leaves to create consistent implementations. Each one of these combinations is a valid customized solution that should be evaluated. All combinations of the remaining tree leaves are evaluated by brute-force exploration.

The exploration methodology consists of two steps and it is presented in Figure 3.8. The inputs of the methodology are the application and hardware constraints. The output is a streaming aggregation implementation with customized software and hardware parameters.

The first step of the methodology aims at the pruning of the design space and the implementation of the DSE. First, the non-applicable options are removed from the design space due to the application and hardware constraints. Then, the streaming aggregation is executed once for each different combination of the decision tree leaves of the design space. For each customization, throughput, latency, memory size and energy consumption

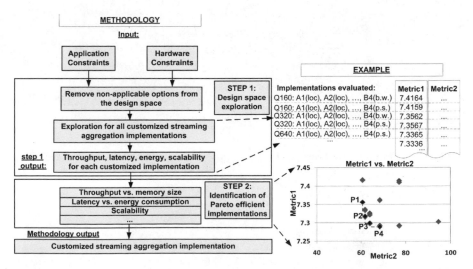

Figure 3.8   Instantiation of the methodology in the context of multiway streaming aggregation.

results are gathered. Scalability is another metric that can be evaluated, in case there is a relatively large number of PEs available. In the second step, the Pareto efficient implementations are identified. The trade-offs that can be performed by customization of the streaming aggregation on an embedded platform are presented in the form of Pareto curves. Developers can select the implementation that is most efficient according to the optimization target.

Figure 3.9 shows some results of the outcome of the methodology for two different embedded platforms: Movidius Myriad2 [14] and Exynos [15]. Figure 3.9(a) and Figure 3.9(c) show the performance of each implementation for throughput vs. required memory size, while the rest show latency vs. energy consumption. The methodology provides a systematic way to find the most effective configuration that adheres to the design constraints. Also, several observations regarding the impact of each design option to the

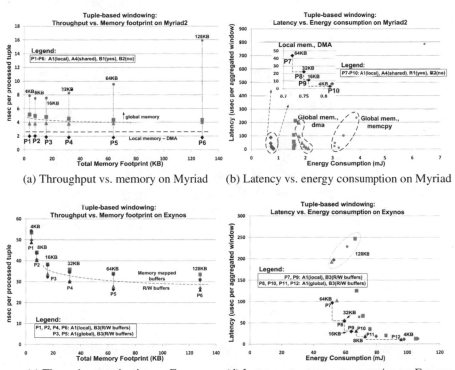

(a) Throughput vs. memory on Myriad   (b) Latency vs. energy consumption on Myriad

(c) Throughput evaluation on Exynos   (d) Latency vs. energy consumption on Exynos

Figure 3.9 Demonstration of tuple-based multiway streaming aggregation on Myriad and Exynos architectures.

evaluation metrics can be reach by the exploration process performed by the implementation of the methodology. For example, by increasing the queue size, energy consumption drops, while latency increases (Figures 3.9(b,d)). More details about the instantiation of the methodology on the context of streaming aggregation can be found in Rosvall and Sander [16].

3.3 Conclusion

In this chapter, we presented a methodology based on DSE that provides a systematic way to effectively explore complex design spaces. The methodology can be applied in various contexts and assist the development of tools that can automate several of its steps. The exploration time is reduced, since the irrelevant configurations are automatically pruned and the extensibility features can be used to further extend the design space with more options.

References

[1] Rosvall, K., and Sander, I. (2014). "A constraint-based design space exploration framework for real-time applications on mpsocs." in *Proceedings of the Conference on Design, Automation & Test in Europe* (Belgium: European Design and Automation Association), 326.

[2] Grüttner, K., Hartmann, P. A., Hylla, K., Rosinger, S., Nebel, W., Herrera, F., et al. (2013). The complex reference framework for hw/sw co-design and power management supporting platform-based design-space exploration. *Microprocess. Microsyst.* 37, 966–980.

[3] Givargis, T., Vahid, F., and Henkel, J. (2001). "System-level exploration for pareto-optimal configurations in parameterized systems-on-a-chip," in *Proceedings of the 2001 IEEE/ACM International Conference on Computer-aided Design*, ICCAD '01 (New York, NY: IEEE) 25–30, Piscataway, NJ, USA.

[4] Ganapathy, S., Karakonstantis, G., Canal, R., and Peter Burg, A. (2014). "Variability-aware design space exploration of embedded memories," in *2014 IEEE 28th Convention of Electrical and Electronics Engineers in Israel (IEEEI)* (New York, NY: IEEE), 1–5.

[5] Palermo, G., Silvano, C., and Zaccaria, V.(2005). Multi-objective design space exploration of embedded systems. *J. Embed. Comput.* 1, 305–316.

[6] Baloukas, C., Risco-Martin, J. L., Atienza, D., Poucet, C., Papadopoulos, L., Mamagkakis, S., et al. (2009). Optimization methodology of

dynamic data structures based on genetic algorithms for multimedia embedded systems. *J. Syst. Softw.*, 82, 590–602.

[7] Xydis, S., Bartzas, A., Anagnostopoulos, I., Soudris, D., and Pekmestzi, K. (2010). "Custom multi-threaded dynamic memory management for multiprocessor system-on-chip platforms," in *2010 International Conference on Embedded Computer Systems (SAMOS)*, 102–109.

[8] Ammar, M., Baklouti, M., and Abid, M. (2016). The performance-energy tradeoff in embedded systems design: a survey of existing design space exploration tools and trends. *Int. J. Comput. Sci. Inform. Secur.* 14, 381.

[9] Jung, C., Rus, S., Railing, B. P., Clark, N., and Pande, S. (2011). "Brainy: effective selection of data structures," in *ACM SIGPLAN Notices* (New York, NY: ACM), Vol. 46, 86–97.

[10] Freescale Semiconductor Inc. (2015). *Freescale I.MX 6 Quad application processors for industrial products data manual.* Available at: http://cache.freescale.com/files/32bit/doc/data_sheet/IMX6SXIEC.pdf

[11] Moloney, D. (2011). 1tops/w software programmable media processor. *HotChips HC23*.

[12] Papadopoulos, L., Walulya, I., Tsigas, P., and Soudris, D. (2016). A systematic methodology for optimization of applications utilizing concurrent data structures. *IEEE Trans. Comput.* 65, 2019–2031.

[13] Cederman, D., Gulisano, V., Nikolakopoulos, Y., Papatriantafilou, M., and Tsigas, P. (2014). "Brief announcement: Concurrent data structures for efficient streaming aggregation," in *Proceedings of the 26th ACM Symposium on Parallelism in Algorithms and Architectures* (New York City, NY: ACM), 76–78.

[14] Barry, B., Brick, C., Connor, F., Donohoe, D., Moloney, D. Richmond, R., et al. (2015). Always-on vision processing unit for mobile applications. *IEEE Micro*, 256–66.

[15] Chung, H., Kang, M., and Cho, H.-D. (2013). Heterogeneous multi-processing solution of exynos 5 octa with ARM® big. LITTLE™ Technology. Available at: https://www.arm.com/files/pdf/Heterogeneous_Multi_Processing_Solution_of_Exynos_5_Octa_with_ARM_bigLITTLE_Technology.pdf

[16] Papadopoulos, L., Soudris, D., Walulya, I., and Tsigas, P. (2016). Customization methodology for implementation of streaming aggregation in embedded systems. *J. Syst. Arch.* 66, 48–60.

4

PReDiCt: A Scenario-based Methodology for Realizing Decision-Making Mechanisms Targeting Cyber-Physical Systems

Nikolaos Zompakis[1], Kostas Siozios[2], Lazaros Papadopoulos[1] and Dimitrios Soudris[1]

[1]School of ECE, National Technical University of Athens, Greece
[2]Department of Physics, Aristotle University of Thessaloniki, Greece

Abstract

As systems continue to evolve they rely less on human decision-making and more on computational intelligence. This trend in conjunction to the available technologies for providing advanced sensing, measurement, process control, and communication lead towards the new field of Cyber-Physical System (CPS). Due to superior performance of these systems it is expected to play a major role in the design and development of future engineering platforms with new capabilities that far exceed today's levels of autonomy, functionality and usability. Although CPS exhibits remarkable characteristics, their design and implementation is a challenging issue, as numerous (heterogeneous) components and services have to be appropriately orchestrated together. The problem of designing efficient decision making mechanisms becomes far more challenging in case the target system has to meet also real-time constraints. Throughout this manuscript, we introduce a novel framework to support the efficient, yet flexible, rapid prototyping of decision-making mechanisms for CPSs. Furthermore, a new toolbox that automates the tasks of our framework was also developed as an open-source software. For evaluation purposes, the proposed solution was applied to a building use case in order to support optimum configuration of smart thermostats. Based on our analysis, we depict that the introduced approach maximizes thermal comfort metric with the minimum possible energy cost, as compared to existing state-of-the-art solutions. Additionally, the proposed decision-making mechanism exhibits remarkable lower computational and storage complexity, as compared to available control mechanisms, making it ideal for being executed onto low-cost embedded devices.

4.1 The PReDiCt Framework

Cyber-physical systems operation is distributed in interconnected processing modules (usually heterogeneous). Each module is a link of a chain that operates in an orchestrated interaction with the whole system. Consequently, the modeling and the design of a CPS component requires a holistic approach that considers the optimization goals of the whole system. Note that due to the combinatorial explosion of the possible configurations of the individual components, only an easily adapted modeling in hardware and software can ensure a feasible design. Toward this direction, throughout this section we introduce a novel framework that supports the modeling and rapid prototyping of CPS modules providing a central decision making mechanism that schedules the whole CPS functionality. The last represents a challenging issue considering that the identification of the CPS functionality can not be implemented with a fully deterministic way due to a variety of unexpected operational factors (external environment).

The modeling key features are: (i) the splitting of the design problem in separate steps, (ii) the exploration of the potential solutions-configurations at design time, and (iii) the application of the optimal solutions at run time. In the scheduling part, by classifying and clustering the possible system situations into scenarios, a run-time manager assigns the required resources improving the resource exploitation compared with a conventional approach based on the worst-case scenario.

Figure 4.1 summarizes the main methodology steps including several implementation layers (modeling, simulation, and hardware implementation), evaluating in a final model the efficiency of the scheduling mechanisms. The deployed methodology is a complete in-the-loop flow, consisting of several chronological steps that verify the functionality correctness of each component and the system as a whole. The left side of our V-like approach includes the definition of the requirements, the examined use case scenarios, the available hardware and software components. In the right side, we have the identification and the cost characterization of each system situation as

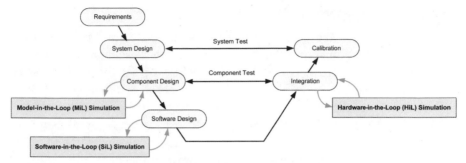

Figure 4.1 The employed V-model approach for designing the decision making mechanisms targeting cyber-physical systems.

well as the realization of the scheduling. The proposed framework also provides the necessary feedback loop to calibrate the system importing changes after the verification step. Below we examine in detail each step.

4.1.1 Step 1: Requirements

In the requirement analysis, we define the system's specifications, considering the potential use case scenarios of the targeted CPS. Note that the output from this task does not include information about the systems development and it refers only to the system's functionality defining the performance and operation characteristics as expected by the users. The goal is to provide the necessary information about the involved modules functionality and interaction and a bounding of the expected cost. In this step various specific-domain languages can be exploited, however, for the scope of our methodology, we prefer the choice of an UML description [13].

4.1.2 Step 2: System Design

The system design step includes the high-level design of the system modules. During this phase, the system is broken up into individual entities (modules), declining the implementation parameters that will be instantiate in hardware or in software. Existing well-known approaches [14, 21] are exploited emphasizing to cover the functionality features that have been described in the requirements description. The high-level design includes: (i) the modules architecture, (ii) the interaction with the external environment, and (iii) the kind and the volume of the related workload.

4.1.3 Step 3: System Modeling

The object of the current step is the deployment of a complete model that simulates the CPS components both as individual entities and as parts of the CPS. The existing available simulation tools address partially the evaluation needs of the CPSs, due to the inherent constraints posed by cyber and physical parts. In order to overcome these issues the proposed framework provides a holistic system rethinking approach that focuses on the way constituent parts interoperate, work over time and function within the context of a larger, evolving system. The current subsection describes the deployed toolbox, depicted schematically at Figure 4.2, in order to support the rapid modeling of a CPS including the scheduling. The introduced toolbox exhibits increased modularity, which is a highly desirable feature as it enables easier framework's upgrade incorporating new functions. In this direction, we have built an open architecture framework that is able to adjust any CPS component-module. Designer can add or replace modules studying their impacts. The idea is to provide a flexible prototyping tool that can provide at an early design phase reliable measurements about the critical characteristics of the final system.

The proposed toolbox relies on a PC-based co-simulation technique, which trade-offs between speed (functional simulation) and accuracy (cycle-

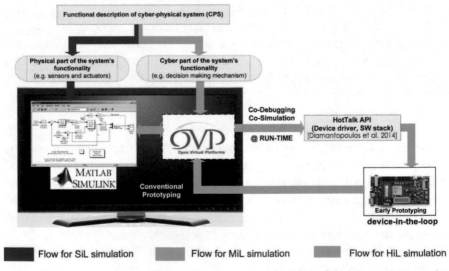

Figure 4.2   Software toolbox for supporting rapid prototyping of decision making mechanisms targeting CPS.

accurate simulation), depending on the design requirements. The incorporated mechanisms for supporting the system modeling (red color flow), simulation (green color flow), and physical design (blue color flow), of a CPS can be though as suitable to alleviate the integration scale problem in RTL simulation, emulation and prototyping environments. On contrast to relevant solutions that are often too complex, slow and expensive, mainly due to the communication link between the host computer and prototyping hardware, the employed approach provides a flexible interface in order to support the co-simulation and co-debugging between the software and hardware parts.

Stating from a high-level representation (previous step), where the functionality of all the system's components (i.e., sensors, actuators, processing elements, etc.) is described, we proceed to the MiL (Model-in-the-Loop) simulation for verifying the system's functionality. All the necessary interactions among modules (though the appropriate interfaces) are also taken into consideration. Different approaches are feasible to be employed for this purpose, while regarding the framework discussed throughout this manuscript we perform this task exploiting Matlab Simulink environment [19]. The competitive advantage of using such a toolset relies mainly to its widely acceptance in research/development community, as well as to the availability of models for composing complex CPS (with increased diversity in terms of component's characteristics).

Next, we proceed to the system's simulation. Even though there are plenty of simulation and design tools that tackle software (SW) and hardware (HW) platforms individually, there are only a few approaches that leverage problems arising in systems that tightly integrate SW and custom HW. This limitation mainly occurs due to the challenges posed by system integration that have to be sufficiently addressed in order to achieve flexible system's simulation. The employed framework considers (without affecting the generality of introduced solution) that the digital part of the CPS is executed onto an embedded processor. For this purpose, the open-source OVP simulator [18] is integrated as part of our toolbox. The increased simulation speed provided by OVPSim ensures that complex systems can be modeled in reasonable amount of time (hundreds of millions of simulated instructions per second). As the OVP models are pre-built, they support fully functional simulation of a complete embedded system. Furthermore, these models are binary-compatible with the simulated HW, hence the developed software can be executed onto the target (final) system without any modifications, resulting to faster software development. We have to notice that OVPSim apart from a wide range of embedded processors can also support custom processing cores (e.g., FPGAs, GPUs, ASICs, etc.) assuming that their equivalent SystemC model is provided. Note that since our framework aims to CPS platforms, the cyber part of the system's functionality (simulated at OVPSim) and the physical part (simulated at Matlab Simulink) have to be simulated as a whole. For this purpose, all the necessary interfaces between these two simulators have also been developed.

Finally, the HiL simulation (depicted with blue color flow) enables the close interaction between SW and HW teams by adopting TLM-SystemC models. Among others such a feature enables as long as new IPs are developed, the HW design team is able to incrementally test these IPs by replacing a functionality of the employed SystemC/TLM model with the equivalent HDL prototype mapped onto FPGA board(s). Regarding the connectivity between the virtual platform (OVP) and the reconfigurable architecture, it is established with our previously published HotTalk API [7].

4.1.4 Step 4: Run-Time Situation (RTS) Definition

The methodology flow continues with the analysis of the functionality introducing the concept of the Run-Time Situation (RTS) [23]. As RTS, we define a piece of system execution with specified and fixed cost dimensions. RTSs are treated as execution units. The whole CPS execution is a sequence of RTSs. The objective of the current step is to identify and characterize the several RTSs at design time. For this purpose is exploited the CPS modeling of the previous step. The first priority is to identify the CPS variables that define the individual RTSs. As a variable is considered each source of variability in the system operation or in the interaction with the environment. The internal system variables represent conditional branches of the system operation. These variables are possible to be tuned at runtime by the users or by a scheduler. The possible tuning choices represent the control knobs of the CPS. Respectively, the external variables represent the environmental factors that influence indirect the operation. For example, in a building that is consisted by a network of sensors and actuators that monitor the inside temperature and tune the air-condition thermostats, the environmental temperature and humidity are external variables that effect the inside tuning decisions.

Considering the alternative values of the internal and external variables we identify for each RTS a combination of tuning knobs. A systematic exploration of all the possible combinations is represented as a tree, as shown in Figure 4.3. The tuning options for each variable are represented as tree leaves. However, not every option is applicable in any context. Some options

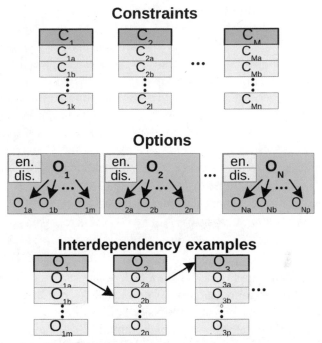

Figure 4.3 Options, constraints, and interdependency examples.

may be non-applicable in specific contexts, being irrelevant or meaningless. Therefore, each option, apart from the values related to the feature or the characteristic that it represents (which are the tree leaves), has one more set of values that are mutually exclusive: enabled and disabled, shown as *en.* and *dis.* in each option of Figure 4.3. To determine the availability of the options in different situations, we identified the constraints that affect the applicability of the options in each specific context. Similarly with the options, the constraints can take different values. Each single combination of constraint values represents a different context in which the system operates. In Figure 4.3, M constraints are shown, each one having a set of possible values that it can take. Therefore, we assigned each single value of each constraint with the enabled or the disabled value of each option. By applying a set of constraint values that represent a specific context, the non-applicable options are disabled (i.e., removed), leaving only the ones that are relevant with the context in which the system is executed. Thus, the total number of identified RTSs can be significantly reduced, since the fact that the non-applicable options are pruned, results in the instantiation of the valid and meaningful RTSs only.

The options are not independent, in the sense that a specific option or options value can make another option or option value non-applicable or meaningless. Therefore, we have defined the set of rules which determine how each option and/or option value affects the applicability of all the others. These rules define the interdependencies between the options. An interdependency is represented with an arrow pointing from option/value A to option/value B. If A is selected during the instantiation of an RTS, then B is non-applicable. For example, in Figure 4.3, the instantiated RTSs for which Option O_1 value is O_1^a cannot take O_2^b value. Also, value O_2^a makes the option O_3 non-applicable. In other words, in all the instantiated RTSs for which O_2 value is O_2^a, O_3 is meaningless or non-applicable.

Based on the above representation, the RTSs can be identified based on the following three step procedure:

1. We represent the system variables in a tree form, for which the leaves are the values that they can take.
2. We identify and apply the constraints. The non-applicable options are removed and the representation is converted from generic to context-specific. Thus, the irrelevant options or option values are pruned.
3. The RTSs can be identified by combining all the remaining tree-leaves by following the interdependencies rules.

Having identified the RTSs the step is concluded with the cost evaluation of the RTSs exploiting the modeling framework of the previous step. This process leads to a Pareto surface [22] of potential exploitation points in a multi-dimensional exploration design space. The Pareto curves allow to unambiguously distinguishing the optimal from the non-optimal points. Thus each RTS is characterized by a Pareto curve that represents its cost impact.

4.1.5 Step 5: Scenario Clustering

After the identification and the evaluation of the response to any situations (RTSs), the system has to be prepared to RTS in specific constraints. The conventional approach to this problem is to design the system reactions based on the worst-case scenario but such an approach is very resource consuming. The key point is to provide a flexible mechanism that adjusts the resource utilization to the running needs. But if this approach is applied for each RTS, the added scheduling overhead will be prohibitively high. The proposed scenario approach is an intermediate solution where representative resource assignments are applied for groups of RTSs instead of individual RTSs. The idea is to encode different RTSs in a reasonable number of system scenarios, leading to significant reduction of the design space. The criterion for the proposed grouping is the cost relevance. Thus, RTSs with similar resource requirements are clustered into the same scenario but without expecting to be equal. This variability leads to a cost overestimation considering that the whole scenario group represented by the worst RTS to ensure any included time constraint and it is expressed as clustering overhead. The clustering overhead is proportional to the average cost variability between the RTSs in the same scenario. Thus, the scenario is a coarse-grain approach (see Figure 4.4) that provides an acceptable trade-off between accurate resource estimation and low complexity scheduling.

Consequently, the targeted system exploits the scenario granularity for implementing a flexible make-decision controller. Such a controller makes use of the scenario costs to schedule the utilization of the available resources. The scope is to achieve the best resource exploitation aiming to the optimal performance without wasting unnecessary resources.

Figure 4.5 presents several resource scheduling approaches highlighting the advantages of the scenario approach. In a case study with three tasks a design approach based on the worst case (Figure 4.5(a)), utilizes a fixed resource budget that corresponds on the worst-case (Task 3). In a fine-grain approach (Figure 4.5(c)), each task corresponds to an individual resource budget while in a scenario approach the two tasks (Task 1 and Task 2) is grouped to one scenario budget. Examining the aforementioned approaches from resource harnessing perspective the fine-grain approach is the optimal as it requires the minimum number of resources (equals to 18), but it imposes three different scheduling configurations (budgets). On the other hand, the worst-case approach requires just one configuration appearing the poorest

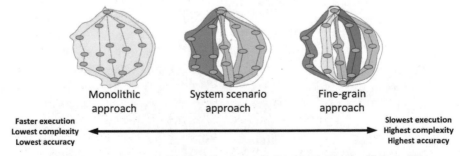

Monolithic   System scenario   Fine-grain
approach   approach   approach

Faster execution   Slowest execution
Lowest complexity ◄─────────────────────► Highest complexity
Lowest accuracy   Highest accuracy

Figure 4.4   Taxonomy of approaches for realizing scheduling mechanisms [8].

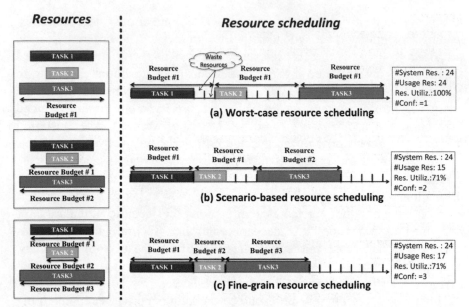

Figure 4.5   Example of resource allocation for the alternative decision making approaches.

resource exploitation (using 24 resources, wasting 6 extra resources). Scenario approach (Figure 4.5(b)) is an intermediate solution, which requires two configurations and an acceptable efficiency in resource exploitation. Note that the system scenario approach is a widely-accepted concept in the embedded domain that handles efficiently design complexity issues with low cost [8].

4.1.6 Step 6: Decision Making

A detection mechanism that identifies the incoming system scenario at runtime and triggers the respecting configuration implements the scheduling mechanism. The scenario detection is implemented based on monitoring the changes of the RTS variables. The implementation cost is proportional to the complexity of the detection (number of the detected scenarios). So heuristic ways, which keep the detection overhead in reasonable levels, are highly desirable. A fundamental trade-off exists between detection and clustering overhead (Step 5). For example a clustering that leads to a few system scenarios with a low detection overhead, it will increase the clustering overhead due to the expected high variability of the RTSs into scenarios. Respectively, an increased number of scenarios will decrease the clustering overhead due to better RTS fitting, but it will increases the detection overhead due to the existence of more scenarios. The detection implementation takes into consideration all these trade-offs [23]. In the context of the current study we instantiate the make-decision scheduler in a FPGA (Field-Programmable Gate Array). The instantiation of the scheduling is the last implementation step that integrates the functionality of the CPS.

4.1.7 Step 7: System Verification

The final step verifies if the system functionality respects the users constraints, revising if it is necessary the design. The verification step provides the Vlike scheme of our approach as depictured in Figure 4.1. The verification includes two criterions the functionality correctness and the time responsiveness. The system is evaluated firstly in the MiL (Model-in-the-Loop) simulation, which consists a mixing of the physical plant model (including mechanical, electrical, thermal, etc.) and the scheduling. Once the MiL simulation verifies the accuracy of our model, we proceed with the software implementation of the digital part of the system's functionality. Note that since the cyber part of a CPS usually is executed onto an embedded processor, all the necessary optimizations for alleviating the complexity and memory requirements have to be applied. Then, the derived software implementation is validated according to system's specifications with the SiL (Software-in-the-Loop) simulation. For this purpose, the source code is simulated with the same physical plant model used during the MiL phase. Although different tests can also be employed for this purpose, it is common to reuse the same test for assuring the equivalence of results (this technique is also known as back-to-back testing). The last phase involves the HiL (Hardware-in-the-Loop) simulation, which is a commonly used technique in the development and test of systems, where part of the functionality is implemented onto hardware components. More thoroughly, HiL simulation provides an effective platform by enabling the real plant to be added in the loop instead of typical (based exclusively on software models) simulation. If the functionality in the HiL simulations is not correct the Steps 2–7 are repeated replacing the out-of-specification components. In case of time violations it is examined if a better scheduling (Steps 5–7) (revising the scenario clustering) can reach acceptable results. Otherwise, the slow modules are replaced to deal with the performance requirements. Thus, the proposed framework is a flexible instrument that provides the opportunity to the designer to intervene in the system deployment, applying *ad hoc* solutions something critical for large-scale systems that it is impossible to predict in advance their behavior.

4.2 Employed Use Case

This section describes the employed CPS, which will be used as a basis for quantifying the efficiency of introduced framework. This system corresponds to a real building consisted of 2 floors/10 offices (located in Chania, Greece), as it is depicted at Figure 4.6 [11]. The employed building uses a number of sensors for monitoring temperature (both inside and outside the building), as well as sunshine and humidity. For our experimentation, we assume (without affecting the generality of introduced solution) that these values are acquired once per 10 min. Furthermore, the building is equipped with photovoltaic arrays in order to minimize the energy consumption. The analysis discussed at the rest of this manuscript considers the problem of operating air-conditioners during the summer period (June, July, and August), in order to cool-climate the rooms. For this purpose, real historical weather data are employed, as they were collected during the 2010 [6]. The objective is to define a solution that takes into consideration both the energy efficiency, as well as the user comfort satisfying level. Specifically, the control input that the user can actuate is the HVAC (heating, ventilating, and air conditioning) set point in each of the 10 rooms. The energy consumed by the HVAC is a nonlinear function which depends on the difference between the set point, the zone temperature, the current external environment conditions, as well as on factors related to the particular unit (e.g., its capacity, efficiency and other construction factors). For this analysis, we assume that the HVACs operate only during the occupancy hours of the building (6 am–9 pm), while energy savings occur from the employed solar panels (depending on the current weather conditions).

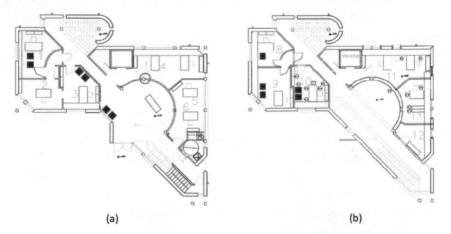

(a)   (b)

Figure 4.6   Floorplans for the (a) ground and (b) first floor of our smart building.

The problem of deciding upon HVAC configuration is a well-established challenge. For decades, building management systems have automated the process of providing just enough energy to heat/cool buildings in order to meet comfort standards. This is a quite challenging problem, where the control system attempts to exploit "as much as it can" the renewable energy so as to reduce the demand for non-renewable energy (coming from the grid) or during time-slots of low-cost tariffs, while maintaining user comfort (i.e., making sure that the building occupants are "satisfied" with the in-building temperature and other thermal conditions). However, apart for this objective, the energy efficiency is also crucial as it highly contributes to an building's sustainability goals, such as tracking and reducing greenhouse gas emissions. But if the data is trapped within the building management system, executive-level decision-makers cannot measure and act on it. The optimization of the operations in buildings is attracting the interest of many researchers, and several approaches and problem formulations have been proposed in the relevant literature. Despite the very impressive recent advances in control system theory, the development of efficient control strategies for Large Scale Systems (LSS) still remains an open and challenging issue [12].

The *"curse-of-dimensionality"*, still haunts control engineering. For instance, despite the significant progress made in optimal nonlinear control theory [3] the existing methods are not, in general, applicable to LSS because of the computational difficulties associated to the solution of the Hamilton-Jacobi partial differential equations. Similarly, Model Predictive Control (MPC) for nonlinear systems, a control approach which has been extensively analyzed and successfully applied in industrial plants during the latest decades [15, 16], faces also dimensionality issues: in most cases, predictive control computations for nonlinear systems amount to numerically solving on line a non-convex high-dimensional mathematical programming problem, whose solution may require a quite formidable computational burden if on line solutions are required. Another family of approaches employ optimization-based schemes to calculate the controller parameters [4]. These approaches require analytical calculation of Jacobian and Hessian matrices, which in LSS is a very time consuming process. Existing simulation-based approaches are not able to efficiently handle systems of large-scale nature as this requires solving a complex optimization problem with hundreds or thousands of states and parameters [17, 20].

Despite the impressive results achieved by the previously discussed control approaches, their increased computational complexity (which usually comes with a demand for excessive amount of storage) makes their realization as part of an embedded system non-realistic. This problem becomes far more challenging at the CPS domain, mainly due to the large scale of these systems. Thus, the rest of the manuscript focuses on applying the introduced framework in order to design and instantiate a scenario-based approach for supporting the decision making at smart thermostats.

In order to describe in more thoroughly the system's architecture, Figure 4.7 presents in UML form the main functionalities performed by the smart thermostats regarding the general form of the employed use case. Specifically, starting by collecting a number of weather-related data (e.g., temperature, humidity, and radiation), as they were acquired by the weather station, it is possible to analyze the efficiency of different thermostat configurations. The results are fed as input to the scenario analysis in order to compute the optimum scenario set in a season basis (winter, spring, summer, and autumn). Then, the employed controller implement the desired policies in order to maximize user's comfort with the minimum energy cost. During this phase, alternative techniques for further energy savings (e.g., building warm-up phase, switch-off heating/cooling in case there is no motion detection) are also applicable. However, these techniques are beyond the scopes of this study, as we focus solely at the decision-making.

The evaluation of derived solutions are quantified with two unrelated metrics, namely the user's comfort satisfying level and the energy management, as it is defined by Equation 4.1 While the energy cost can be easily quantified, different methods are available for measuring the thermal comfort. In this work we concentrate on the thermal comfort model developed by Fanger [1] which evaluates the PPD people in a room. Fanger proposed an equation for thermal comfort that relates environmental and physiological factors with the

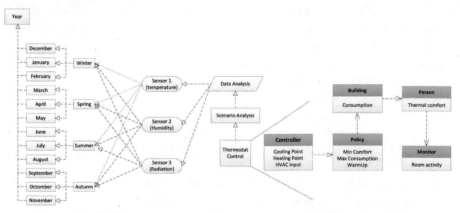

Figure 4.7   UML representation of the smart thermostats problem formulation.

thermal sensation. Note that Fanger equation is usually solved numerically, while the majority of existing building simulators directly provide the corresponding PPD Fanger index (in %), using charts and diagrams based on statistical data from which optimal comfort conditions can be read given a knowledge of metabolic rate and clothing insulation. Finally, the value of k metric defines the importance of optimizing either the energy bill, or the user comfort. Note that, the two terms of the total score are competitive parameters; thus it is not feasible both of them to be optimized simultaneously. For instance, in order to improve the satisfaction for occupants, it is necessary to perform room cooling/heating (spend additional energy). On contrast, the employed decision making approach (similar to the majority of existing control techniques) aim to minimize the total energy with a penalty at the occupants comfort.

$$\text{Cost}(t) = k \times \text{Energy}(t) + (1 - k) \times \text{Comfort}(t) \qquad (4.1)$$

4.2.1 Applying the Proposed Framework

The design of a decision-making mechanism for the target case study exploits the multi-level modeling of the aforementioned PReDict framework. Methodology starts with the identification of the system variability. As system, we assume the overall CPS, while a system variable corresponds to each source of variation that affects the cooling/heating of the building. The source of variation can occur either internally (by the building) or externally (by the environment). Based on this criterion we can distinguish two different kinds of system variables, the *internal* and the *external* variables. The internals are parts of the close system (building) while the externals represent the interactions with the external environment. The system variables, as we have already referred at methodology, can be further distinguished in two extra categories; at *condition variables* and *control knobs*. Condition variables define the dynamic situations under which the system comes to operate, while the control knobs correspond to the operation modes with which the system responses. For example, in our case study outside temperature, humidity, and radiation are the condition variables while the configurations for the building thermostats are the control knobs. Many of the condition variables can be differentiated with an un-expected way and the full knowledge of their behavior usually is a very complicated (non-realistic) problem. The proposed scenario-based approach concentrates on the variables with the most significant impact (e.g., specific ranges of the temperature or humidity), which are known as

RTS parameters. To reduce further the design effort, dependencies between these variables (e.g., a variable can fully dependents from another variable this variables is ignored) are explored. For example for specific temperature and radiation values, humanity can not range over specific levels. Thus, a classification of the system variables is implemented based on their impact and their dependencies to distinguish the RTS parameters. These parameters define an exploration space with all the potential RTSs.

Having modeled all the architectural properties of the target building (as they were discussed at Figure 4.6) the optimization effort focuses on minimizing the cost function discussed at Equation (4.1) which takes into account both the energy consumption, as well as the user comfort. Note that as we mentioned, the building is equipped with photovoltaic panels. Thus, the energy cost $\text{Energy}(t)$ depends both on the power from the grid $\text{Energy}_{\text{cons}}(t)$, as well as the potential savings from solar panels $\text{Energy}_{\text{sav}}(t)$ according to Equation (4.2). Consequently, an efficient decision-making approach is not a trivial determination of one cost metric, but it leads to a Pareto surface of potential exploitation points in the multi-dimensional exploration space. Each cost metric adds an addition dimension in the overall cost space. Each system configuration represents a distinct cost trade-off that is called as design space point, while the objective is to extract the optimal design space points based on Pareto optimality [22].

$$
\text{Energy}(t) =
\begin{cases}
\Big(\text{Energy}_{\text{cons}}(t) - \text{Energy}_{\text{sav}}(t)\Big), & \text{if } \text{Energy}_{\text{cons}}(t) \geq \text{Energy}_{\text{sav}}(t) \\
0 & , \text{if } \text{Energy}_{\text{cons}}(t) < \text{Energy}_{\text{sav}}(t)
\end{cases}
\tag{4.2}
$$

4.3 Experimental Results

This section summarizes the results for our experimentation regarding the usecase described at previous section. Without affecting the generality of our evaluation procedure, we set the desired temperature at each room once per 10 minute, since this period is a typical approach for configuring HVAC systems. For comparison purposes, the introduced scenario-based decision making is evaluated against the static temperature configuration, as well as to two well-established control techniques. For sake of completeness, our experimentation is performed with historical data for the summer 2010 at

Chania city (the third week per month). Figure 4.8 visualizes the variation of outside temperature, humidity and radiation for our experimentation, as it was acquired by the employed weather stations. Based on this figure we might conclude that all the three parameter vary a lot for the studied period; hence, the conclusions derived about the efficiency of introduced control mechanism can be though as representative for a general-purpose smart thermostat.

Specifically, the variation for temperature ranges between 20° and 35° while the corresponding ranges for humidity and radiation are 7.4–17.2×10^{-3} kg/m and 0–886 KW/m, respectively. For each of these combinations, typically a different configuration setup has to be computed for the alternative decision making methods. Thus, effectiveness also in term of execution run-time should also be taken into consideration, as the smart thermostats rely typically on low-power embedded processors and/or

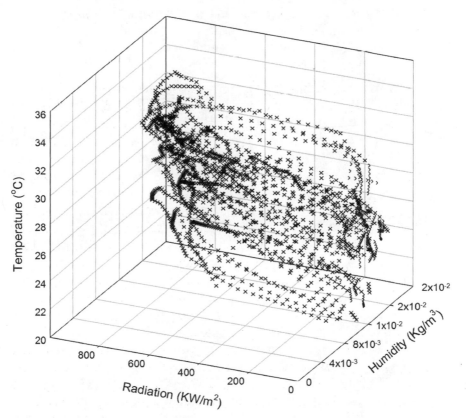

Figure 4.8 Representative variation of temperature, humidity, and radiation parameters for our experimentation.

micro-controllers. As we have discussed previously, the two metrics for quantifying the quality of solutions (configurations for the thermostats) are the thermal comfort and the energy consumption. Both of these metrics are computed per day for the working hours (6:00 am–21:00 pm), where people present and HVAC are operating. Also, during our experimentation, by default the employed decision making mechanism pays effort to optimize simultaneously the two cost factors (k=0.50).

For visualization purposes Figures 4.9 and 4.10 plot the variation of thermal comfort and energy dissipation parameters, respectively, regarding a representative day during July. This analysis indicates that our approach provides the optimal configuration of smart thermostats regarding the thermal comfort metric, since it results to the minimum Fanger value among the alternative strategies during the whole day. Regarding the spike at the beginning of building's operational period (6:00 am), which is common for all the studied configuration, it occurs mainly due to the people entrance in a "closed" building; thus, the conditions cannot be though as optimal for all the approaches. Apart from the thermal comfort, we are equally interested to reduce energy cost.

In order to visualize this parameter, Figure 4.11 plots the energy dissipation for the alternative approaches, as well as the energy savings achieved from the photovoltaic panels (depicted with dotted line). The information about the energy savings achieved with photovoltaic panels is taken into

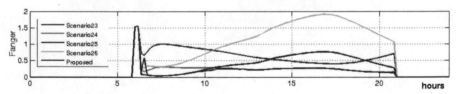

Figure 4.9 Variation of thermal comfort achieved with alternative temperature configuration strategies regarding a representative day of July.

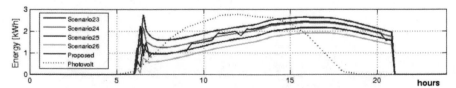

Figure 4.10 Variation of energy consumption achieved with alternative temperature configuration strategies regarding a representative day of July.

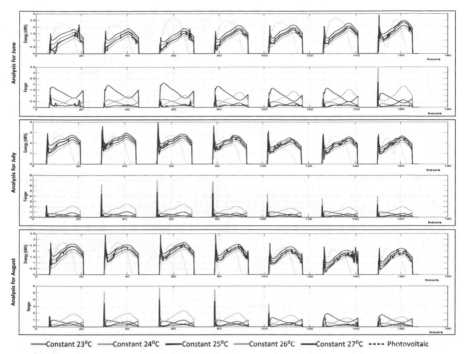

Figure 4.11 Analysis in terms of energy consumption and thermal comfort for the alternative temperature configuration strategies.

consideration in order to further improve thermal comfort (e.g., by modifying the aggressiveness of building cooling during the day period). The results summarized at this figure indicate that our decision making leads to energy cost similar to those occurred by using constant 24° and 25°; however, as we discussed previously, we achieved better thermal comfort as compared to the ones achieved with the static configurations.

The corresponding results regarding the variation of thermal comfort and energy consumption among alternative temperature configurations strategies for the entire summer (3 months period) are summarized at Figure 4.11. For demonstration purposes (and without affecting the generality of introduced solution), this analysis is performed only at the 3rd week/month. Note that this analysis can also be performed for additional weeks (e.g., the entire summer period), however the results cannot be easily visualized at a similar graph (due to the selected fine-grain granularity).

A number of conclusions might be derived based on this analysis. Among others, there are a lot of similarities both for the distribution of thermal comfort, as well as the energy cost, between days, which is expected due to the summer period. Furthermore, we can identify time periods per day (e.g., morning and afternoon), where there are increased values for thermal comfort, while at the noon it seems that energy consumption is maximized (since there are negligible savings from photovoltaic panels) with a corresponding improvement at thermal comfort. Although the previously mentioned conclusions are exhibited during the entire summer, rarely they are taken into consideration from existing temperature control strategies. On contrast, the introduced decision-making approach exploits these similarities in order to increase design flexibility by clustering (at a few scenarios) RTSs with close cost metrics.

In order to evaluate more thoroughly the efficiency of the alternative temperature configuration strategies Figure 4.12 plots, the energy consumption (KhW) and the Predicted Percentage Dissatisfied (PPD) for the entire summer period. Furthermore, three different instantiations of the introduced decision making mechanism are quantified, which correspond to solutions retrieved with k=0.50 (equally important the optimization of thermal comfort and energy consumption), k=0.25 (emphasis on improving thermal comfort), and k=0.75 (emphasis on improving energy dissipation). Based on this analysis, it is clear that higher set points lead to less energy consumption, since the cooling is not that intense. Regarding the proposed decision-making mechanisms, it results always to better solutions in term of PPD, while the energy consumption is similar to the average energy cost for the constant temperature configurations. Also, it is well-worth to mention that the overall picture for the three instances of proposed approach is almost constant for the entire summer period, which proves that our solution takes into consideration the weather data (as it is acquired by the sensors of weather station) in order to configure appropriately building's cooling strategy.

With respect to quantify the overall efficiency of decision-making mechanism (by taking into consideration both energy cost and thermal comfort), Figure 4.13 plots the average cost values of different solutions (for the 3-month experiment). For demonstration purposes, the vertical axis at this figure is given in normalized manner over the corresponding solutions retrieved with the static configuration of thermostats with 24° (referred as Constant 24°). This analysis shows that our framework leads to higher quality solutions, as our typical instantiation (k=0.50) achieves to reduce further the optimum cost retrieved with static configurations (24°) by 15%.

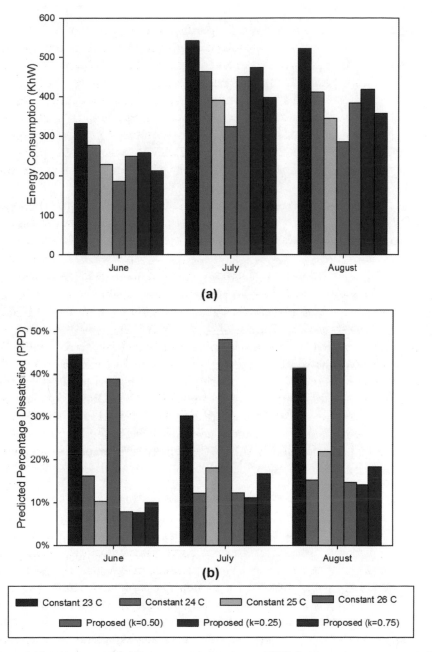

Figure 4.12 Average results for energy consumption and PPD for the entire summer period.

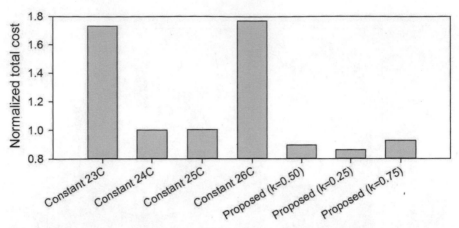

Figure 4.13   Total cost (plotted in normalized manner versus the "Constant 24°" configuration) regarding the results depicted at Figure 4.12.

The efficiency of our introduced decision-making mechanism (assuming equal effort for optimizing the energy cost and thermal comfort) is also quantified against two widely accepted techniques for supporting sophisticated control in HVAC operations. Both of these techniques, named Fmincon [5] and Pattern Search [2] are implemented in the Optimization Toolbox of Matlab, while for the scopes of our experimentation they form an open-loop optimization procedure. Similar to our case study, these control approaches optimize a vector that corresponds to the 3rd week of June (630 values). The results from this analysis are summarized at Table 4.1. The results from this analysis indicate that the Fmincon control achieves superior performance among the studied solution, but imposes extremely large amount of execution time (note that each iteration of EnergyPlus at Matlab need about 2–3 min at a

Table 4.1   Comparison (average values) improvement of decision making techniques as compared to constant temperature values

Approach	Imrovement wrt Scenario 24°C	Improvement wrt Scenario 25°C	Iterations
Fmincon	37.7%	16%	≈ 5,000 Matlab iterations
Pattern Search	26.4%	1%	≈ 5,000 Matlab iterations
Proposed (k=0.50)	27.9%	4%	N/A[1]

---

[1]Since our instantiation of decision making is not implemented at Matlab, we cannot quantify this metric. Details about the performance metrics of our implementation can be found in Subsection 4.3.1.

recent desktop PC. Thus, although such a technique achieves superior performance, its computational complexity is not affordable for being implemented as part of an embedded system found in smart thermostats. The difficultly of computing optimal configurations for the thermostats is also depicted by studying the results retrieved with the second well-established approach (Pattern Search). More thoroughly, this control technique leads to solutions with similar quality as compared to those retrieved from our framework.

Although the previously mentioned analysis indicates that existing control techniques might improve the overall quality of the retrieved solutions, it is well-worth to analyze the impact of different decisions making from the alternative approaches in terms of thermal comfort and energy consumption. Toward this direction, Figure 4.14 plots the temperature variation between the setpoints at smart thermostats and the outside temperature for the entire period of our experimentation. This analysis can be though as a direct extension of Fanger metric, because it highly affects the occupant's thermal comfort. These results highlight that the alternative decision-making approaches perform almost identical, since their average variation is less than $0.3°$; thus it is not expected to be noticed by people. Similarly, Figure 4.15 plots the variation of energy consumption for the same experiment. Note that since this analysis refers to energy consumption, we are not interested about peak values but for the area surrounded by each curve. Based on

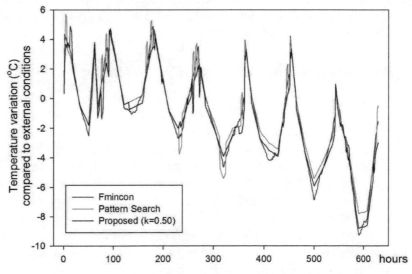

Figure 4.14 Variation of temperature at smart thermostats vs. outside temperature.

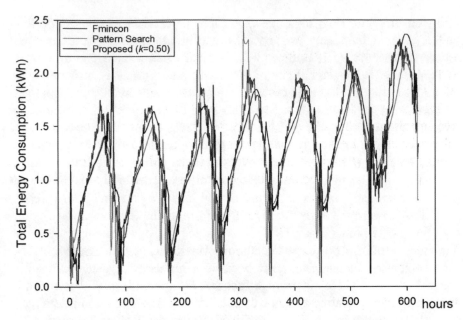

Figure 4.15   Variation of energy consumption for alternative control techniques.

this figure it is clear that our introduced scenario-based approach leads to comparable energy consumption, as compared to the results retrieved for the existing (Fmincon) approach. Consequently, we can claim that our framework achieves almost identical performance (at system-level) in comparison with the existing complex control approaches.

4.3.1 Hardware Implementation of Decision Making

The most computationally intensive kernel of our system's algorithms was also implemented onto an FPGA. Typically, embedded FPGAs are low-cost and thus they exhibit limited capacity and performance compared to regular ones. The selected FPGA device is based on projections regarding the expected capabilities of smart thermostats in the next 5 years, as they can be concluded in accordance with various roadmaps [9, 10]. Such an analysis indicated that by that time, FPGAs qualified for these products will offer hardware resources and capabilities similar to today's Xilinx Spartan-3 family – a low cost device that provided sufficient enough resources for realizing the majority of functionalities expected to be found at such products in the near future. However, for shake of completeness, during our experimentation we also quantify the efficiency of implementing the same decision-making mechanism onto most recent devices (e.g., Xilinx Zynq-7000). The system's architecture is hierarchically structured in three levels. Specifically, Level 0 corresponds to the building blocks of the system (such as sensors and actuators), and it is implemented on top of the Matlab Simulink [19]. The components in Level 1 are the C/C++ classes implementing the functionality of the higher-level elements in Level 0. This development emphasized on realizing in efficient manner the decision making algorithms. Finally, Level 2 corresponds to those computationally intensive kernels that are mapped on the FPGA. The final definition of these kernels was performed as the outcome of the co-design and partitioning procedures. As a result, the used algorithms were partitioned into a standard processing software component and a VHDL model component. The former runs on an embedded processor executed onto OVP simulator [18] exploiting the features provided by the operating system, whereas the latter is implemented in programmable hardware with FPGA technology. Table 4.2 outlines the demand for resources in order to realize the scenario-based decision making, as well as a number of performance metrics. The results indicate the efficiency of our solution, since it can be executed at a few cycles, in comparison to the $\approx$ 5000 iterations of Matlab for relevant approaches, as they were discussed at Table 4.1. Each iteration of Matlab takes around 2–3 min on a recent desktop PC (Intel i5-4590 CPU@3.3GHz with 8 GB RAM. Thus, the introduced control approach can be applied either to larger scale problems (e.g. control additional rooms, take into consideration parameters such as air velocity, etc), or it can be executed at a finer time step granularity.

Table 4.2 Comparison (average values) improvement of decision making techniques as compared to constant temperature values

Parameter	Spartan-3E			Virtex-4			Zynq-7000		
Number of Flip Flops	712	4,896	14%	744	10,944	6%	618	35,200	1%
Number of LUTs	3,058	4,896	62%	3,154	10,944	28%	1,690	17,600	9%
Number of occupied Slices	1,837	2,448	75%	1,905	5,472	34%	550	4,400	12%
Clock (MHz)	137.4			247			263		
Power (mWatt)	52			167			65		

4.4 Conclusion

A novel methodology for supporting the rapid design complex CPSs, was introduced. This methodology is software-supported by a new toolbox that automates different tasks of the system's modeling, simulation, and development (both at hardware and software levels). In order to quantify the efficiency of introduced framework, it was employed to design the decision-making mechanism for a CPS. Specifically, by enabling the run-time configuration (depending on inputs acquired from sensors at real-time) of smart thermostats in a large scale building, we shown that it is feasible to achieve simultaneously both energy savings, as well as to maximize the occupant's thermal comfort. Additionally, we prove the superior of our solution in respect to existing control techniques, as the introduced mechanism exhibits remarkable lower computational complexity with an almost negligible penalty at quality of derived configurations.

References

[1] ASHRAE. (2017) *Ansi/ashrae standard 55–2004: Thermal environmental conditions for human occupancy.* Available at: https://www.ashrae.org/File (accessed April 30, 2017].

[2] Audet, C., and Dennis, J. E. Jr. (2002) Analysis of generalized pattern searches. *SIAM J. Optim.* 13, 889–903.

[3] Basar, T., and Bernhard, P. (2008). *H-infinity optimal control and related minimax design problems: a dynamic game approach.* Berlin: Springer.

[4] Bertsekas, D. P. (1999). *Nonlinear Programming.* Belmont: Athena Scientific.

[5] Byrd, R. H., Gilbert, J. C., and Nocedal, J. (2000). A trust region method based on interior point techniques for nonlinear programming. *Math. Program.* 89, 149–185.

[6] Diamantopoulos, D., Sotiriou-Xanthopoulos, E., Siozios, K., Econo-makos, G., and Soudris, D. (2014). Plug&chip: A framework for supporting rapid prototyping of 3d hybrid virtual socs. *ACM Trans. Embed. Comput. Syst.* 13, 168: 1–168: 25.

[7] EnergyPlus. (2017). *Weather data sources for energyplus framework.* Available at: http://apps1.eere.energy.gov/buildings/energy plus/weatherdata_sources.cfm/ [accessed April 30, 2017].

[8] Gheorghita, S. V., Basten, T., and Corporaal, H. (2008). Application scenarios in streaming-oriented embedded-system design. *IEEE Design Test Comput.* 25, 581–589.

[9] HiPEAC. (2015) *Hipeac vision.* Available at: https://www.hipeac. net/assets/public/publications/vision/hipeac-vision-2015_Dq0boL8.pdf [accessed April 30, 2017].

[10] ITRS. (2013). *International technology roadmap for semiconductors.* [accessed April 30, 2017].

[11] Korkas, C. D., Baldi, S., Michailidis, I., and Kosmatopoulos, E. B. (2015). Intelligent energy and thermal comfort management in grid-connected microgrids with heterogeneous occupancy schedule. *Appl. Energy* 149, 194 – 203.

[12] Kosmatopoulos, E. B. (2008).Adaptive control design based on adaptive optimization principles. *IEEE Trans. Automatic Control* 53, 2680–2685.

[13] Kukkala, P., Riihimaki, J., Hannikainen, M., Hamalainen, T. D., and Kronlof, K. (2005). "Uml 2.0 profile for embedded system design," in *Design, Automation and Test in Europe*, Vol. 2 (New York, NY; IEEE) 710–715.

[14] Li, Y., Callahan, T., Darnell, E., Harr, R., Kurkure, U., and Stockwood, J. (2000). "Hardware-software co-design of embedded reconfigurable architectures," in *Proceedings 37th Design Automation Conference* (New York, NY; IEEE), 507–512.

[15] Magni, L., De Nicolao, G., Magnani, L., and Scattolini, R. (2001). A stabilizing model-based predictive control algorithm for nonlinear systems. *Automatica*, 37, 1351–1362.

[16] Mayne,D. Q., Rawlings, J. B., Rao, C. V., andScokaert, P. O. M. (2000). Survey constrained model predictive control: Stability and optimality. *Automatica*, 36, 789–814.

[17] Nghiem, T. X., and Pappas, G. J. (2011). "Receding-horizon supervisory control of green buildings" in *Proceedings of the 2011 American Control Conference* (IEEE: New York, NY), 4416–4421, June 2011.

[18] OVP. (2017). *Open virtual platforms* (ovp). Available at: http://www.ovpworld.org [accessed-April 30, 2017].

[19] Simulink. (2017). *Simulation and model-based design.* Available at: http://www.mathworks.com/products/simulink/ [accessed April 30, 2017].

[20] Trcka, M., Hensen, J., and Wetter, M. (2009). Co-simulation of innovative integrated hvac systems in buildings. *J. Build. Perform. Simul.* 2, 209–230.

[21] Yen, T.-Y., and Wolf, W. (1996). *Hardware-Software Co-Synthesis of Distributed Embedded Systems*. Norwell, MA: Kluwer Academic Publishers.

[22] Yukish, M. A. (2004). *Algorithms to Identify Pareto Points in Multidimensional Data Sets*. Ph.D. thesis, AAI3148694, The Pennsylvania State University, State College, PA

[23] Zompakis, N., Papanikolaou, A., Praveen, R., Soudris, D. and Catthoor, F. (2011). "Enabling efficient system configurations for dynamic wireless baseband engines using system scenarios," in *2011 IEEE Workshop on Signal Processing Systems (SiPS)* (IEEE: New York, NY), 305–310.

5

Studying Fault Tolerance Aspects

Kostas Siozios

Department of Physics, Aristotle University of Thessaloniki, Greece

Abstract

The application domain of cyber-physical systems includes various mission-critical systems, where the system's security and reliability aspects are crucial. Towards this direction, proper mechanisms that are able to handle the challenges posed by zero-faults (or near zero-faults) specifications have been studied. Among others, these mechanisms affect the system's decision-making, which usually dominates the performance of overall CPS architecture. In this chapter, we study aspects related to the fault-tolerance of architecture's components. Although emphasis is given at the hardware infrastructure of these systems, similar (or relevant) techniques are also applicable at software level.

5.1 Introduction

For the last three decades, the microelectronic industry has benefited enormously from the MOSFET miniaturization. The shrinking of transistors to dimensions below 100 nm enables hundreds of millions transistors to be placed on a single chip. The increased functionality and reduced cost of large variety of integrated circuits and systems has brought its own benefit to the end users and above all the semiconductor industry. A low cost of manufacturing, increased speed of data transfer, computer processing power, and the ability to accomplish multiple tasks simultaneously are some of the major advantages gained as a result of transistor scaling. For instance, the majority of consumer electronics nowadays are formed by an excessive number of transistors, without increasing the fabrication cost (and thus, product's cost).

The wide availability of cheap devices however, in conjunction to the variability problems due to technology shrinking, highlights the importance for novel methodologies that are able to guarantee reliable system's execution even in the case where the system is formed by unreliable components. As outcome of this research there are numerous solutions, both at software and hardware level, that can achieve fault masking. Since not all the products exhibit similar requirements for fault tolerance (e.g., mission-critical applications, consumer products, etc.), application-domain, or application-specific fault-tolerant mechanisms, were also introduced. Among others these mechanisms provide an affordable trade-off between the desired level of fault coverage and its overhead (in terms of computational complexity, silicon area, design overhead, etc.).

Another problem is that, although designers do their best to have all the hardware defects and software bugs cleaned out of the system before it goes on the market, this is not always possible. It is inevitable that some unexpected environmental factor is not taken into account, or some potential user mistakes are not foreseen. Thus, even in the unlikely case that a system is designed and implemented perfectly, faults are likely to be caused by situations outside the control of the designers.

The challenge for improving the quality-of-service (QoS) for embedded systems is also inline to the CPS domain. As we have already mentioned, these systems are formed by components that deal with sensing, computation and actuating. All of these tasks are candidate for failures; hence, fault tolerant techniques can improve the overall system's availability. Additionally, as the CPS usually refers to a system that are applied into a wide region, and hence the communication link between individual CPS, or among CPS's components (e.g., sensing, computational and actuators are not present in the same region) has also to be protected against upsets.

The majority of CPS platforms nowadays are somehow related to industrial machine and processes. Thus, the complexity of these tasks becoming increasingly difficult and almost unmanageable using conventional techniques. Therefore, in the past decade, intense research was dedicated to find alternative solutions using methods that are based on mirror human reasoning. These approaches are inspired among others from nature in order to cope with the need for adaptation of the diagnostic methodology to the inherent changes occurring in the diagnosed process. Indeed, one of the main current trends in solving problems in manufacturing industry is developing fault-tolerant control schemes. Fault-tolerant control is concerned with making the controlled system able to maintain control objectives, despite the occurrence of a fault. Hence, fault diagnosis represents the main ingredient of a fault-tolerant control system. Diagnosing the faults that occurred in a system permits triggering control mechanisms to keep a plant working sufficiently well until the necessary maintenance may be performed. In practice, this feature results in a significant improvement in industrial plant safety, productivity, and time in service.

The automatic diagnosis requires the ability to identify the symptoms automatically and map them to their causes as well as, eventually, to prescribe solutions for repairing/restoring the good functionality of the device, machine or plant. Some methods can prove suitable for certain systems while being totally inappropriate for others. Among others this automation emphasizes on the reduction of computational complexity in order to derive system's customization. For this purpose, research community has introduced a number of heuristic approaches that solve complex problems in reasonable execution time have been derived. The heuristic algorithms and the computational intelligence techniques (e.g., fuzzy techniques, artificial neural networks, genetic algorithms, etc.) are also useful in order to overcome problems related to operating environment of CPS, such as their high nonlinearity, noisy signals, and uncertainty.

More precisely, computational intelligence attempts to emulate human and biological reasoning, decision-making, learning and optimization via a series of techniques that mirror the adaptive evolutionary nature of living beings. Such techniques can be either used individually or combined into more complex hybrid methodologies, resulting in systems with enhanced capabilities, e.g., the same system can benefit from the decision-making under uncertainty enabled by fuzzy logic as well as from learning and adaptation that neural networks provide, or from the evolutionary optimization inherent in genetic algorithms.

Another approach in order to overcome the challenge of increased computational complexity is through the development of distributed fault diagnosis methodologies. The main idea of this approach relies on partitioning the monitored system in subsystems having a reasonable complexity level and, then, to successfully apply state-of-the-art methodologies on each one of them. The global diagnosis of the system is going to be based on all these local diagnosis processes. Such a divide-and-conquer strategy overcomes the system's complexity drawback, resulting to an affordable solution (depending on the system's quality-of-service QoS metrics) in reasonable execution run-time. Additionally, implementing the local diagnosis processes using computational intelligence methodologies retains their ability to treat the local nonlinearities, noise and uncertainty.

This chapter introduces the problems of faults at decision-making mechanisms found in cyber-physical systems and describes a number of techniques that can be used for their masking. Therefore, the techniques presented are applicable to a variety of products, devices, and subsystems.

5.2 Definition of Faults and Fault-Tolerance

A *fault* represents an unexpected change of system function, although it may not lead to a physical failure. Similarly, the term failure indicates a serious breakdown of a system component, or function, that has as a consequence a significantly deviated behavior of the whole system. The term fault rather indicates a malfunction that does not affect significantly the normal behavior of the system. An *incipient (soft) fault* represents a small and often slowly developing continuous fault. Its effects on the system are in the beginning almost unnoticeable. A fault is called *hard or abrupt* if its effects on the system are larger and bring the system very close to the limit of acceptable behavior.

System failure occurs when, due to one or more faults, the system is no longer able to implement the function for which it was designed. For instance, it is possible after a fault, a system not to have the necessary amount of hardware resources in order to complete a task (i.e., the amount of memory is not sufficient to perform a computation). Alternatively, a system may simply lack the resources to complete a task before a given deadline.

The three fundamental terms in the present context are fault, error, and failure. While these terms are often used synonymously in the literature, they are not identical but rather related by the following cause-and-effect relationship: *faults are the cause of errors, and errors are the cause of failures*. A different view of the cause-and-effect relationship mentioned above is the classification into different abstraction levels at which faults, errors and failures occur. More specifically, failures occur in the external level, errors occur in the informational level, while faults occur in the physical level.

In order to prevent, or at least alleviate, the consequences of faults, a number of mechanisms, also known as *fault–tolerant solutions*, have been proposed. The term fault-tolerant corresponds to a design able to continue its operation, possibly at a reduced level, rather than failing completely, when some part of the system fails. A fault-tolerant system should be able to handle faults in individual hardware or software components, power failures, or other kinds of unexpected problems and still meet its specification.

Fault tolerance is necessary because it is practically impossible to build a perfect system. The fundamental problem is that, as the complexity of a system grows, its reliability drastically decreases, unless compensatory measures are taken. For instance, if the reliability of individual components is 99.99%, then the reliability of a system consisting of 100 non-redundant components is 99.01%, whereas the reliability of a system consisting of 10,000 non-redundant components is just 36.79%. Such a low reliability is unacceptable in most applications. If a 99% reliability is required for a 10,000-component system, individual components with a reliability of at least 99.999% should be used, implying a sharp increase in cost.

Even though fault tolerance could be though as a pre-request for the cyber-physical architectures, the excessive mitigation cost makes it affordable only for mission critical systems. However, there are numerous applications that can afford lower fault coverage for significantly reduced mitigation cost. Up to now, a number of architectures and design methodologies able to provide non-distributed device operation have been proposed at different levels of abstraction. Specifically, in literature, there are two mainstream approaches for designing fault-tolerant systems. The first of them deals with the design of new hardware elements, which are fault tolerant enabled, whereas the desired fault masking at the second approach is provided at software level with the usage of specialized software tools. Both approaches exhibit advantages and disadvantages. The (re-)designed hardware blocks can either replace the existing components at conventional CPS, or new fault tolerant CPS architectures can be designed to improve robustness. The drawback of applying such a strategy is the increased design complexity, while the derived CPS provides also a static (defined at fabrication time) fault tolerant mechanism. Typical instantiations of this approach involve the usage of spare logic and interconnection resources. On the other hand, the software-based fault masking combines the required dependability level with the low-cost of commodity CPS platforms. However, the software-based fault tolerant systems assume that the designer is responsible for protecting the target application. Since this approach does not impose any hardware modifications, it is widely accepted for research and product development.

Critical role to the design of reliable CPS, where their functionality is not disturbed by upsets, is given to the fault-tolerant control system. This system is a controlled system that continues to operate acceptably following faults in the system or in the controller. An important feature of such a system is automatic reconfiguration, once a malfunction is detected and isolated.

Fault diagnosis contribution to such a fault-tolerant control system is detection and isolation of faults in order to decide how to perform reconfiguration.

fault diagnosis task consists of two main stages: *residual generation* and *decision-making* [5] (Figure 5.1). Residual generation is a procedure for extracting fault symptoms from the system, using available input and output information. A residual generator represents an algorithm used to generate residuals [6]. Decision-making represents examining the residual signals in order to establish if a fault occurred and isolate the fault.

5.3 Overview of Wear-out Mechanisms

Defects at CPS platforms are tightly coupled to the operating conditions. As manufacturing processes scale, permanent faults occur more frequently due to wear out because of increased strain on ever smaller transistors and wires. Wear out is a time-dependent process, whereby over the course of normal operation the integrity of a portion of a device degrades and eventually fails to behave as originally intended, resulting in a permanent wear out

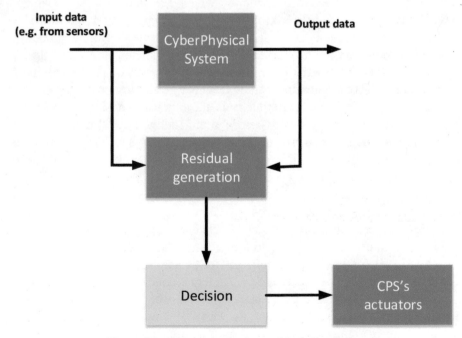

Figure 5.1   The two main stages of fault diagnosis.

induced fault. More specifically, as the power density increased over the last decades due to the technology scaling, it had a consequence that also total power/energy consumption was raising, which in turn leads to higher on-chip temperature values. This has a direct impact on the system's reliability, since the failure rates rise exponentially with the temperature increase.

There can be distinguished three periods of device reliability, as they are depicted in Figure 5.2. In the very beginning during the burn-in period and in the end during the wear out period, the failure rates are high enough. On the other hand, during normal working period, the failure rate remains constant on a rather low level. Wear out phenomenon at digital part of CPS platforms occurs for a variety of reasons and in a variety of ways. Important wear out mechanisms include [7]:

- *Negative Bias Temperature Instability* (NBTI), which is the degradation of transistor performance as charge is implanted in the gate, resulting in timing failures;
- *Time-Dependent Dielectric Breakdown* (TDDB) represents the destruction of gate oxide that occurs when sufficient charge is implanted in the gate to result in a junction-gate short, due to the high-electric field strength;

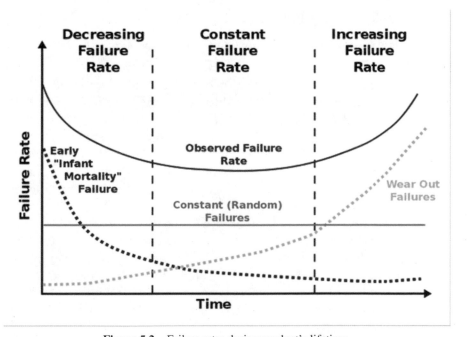

Figure 5.2 Failure rates during product's lifetime.

- *Electro-Migration* (EM) models the movement of metal atoms away from where they had been deposited, induced by the repeated collision of high-energy electrons, eventually resulting in short- or open-circuits;
- *Thermal Cycling* (TC) affects the damage accumulated as a result of uneven expansion and contraction of different parts of the system due to uneven heating and cooling of the system;
- *Stress Migration* (SM) is the movement of metal atoms (much like EM), induced by the uneven expansion and contraction of different materials in the system.

The previously mentioned failure mechanisms have a strong dependence on operating conditions, and more specifically on the temperature values. Since the on-chip temperature increases with technology scaling, the wear-out problem is expected to become far more important in the close future [8]. Additionally, due to leakage power, both the supply, as well as the threshold voltage, no longer scale ideally. As a consequence, power density is increasing, and, therefore, so is system temperature. Among others, the increase at temperature values also imposes higher mean-time-to-failure (MTTF) rates because the EM, TDDB, SM, and NBTI decreases exponentially, while the MTTF due to TC decreases proportionally to $(1/T)^{2.35}$ [7].

5.4 Classication of Faults

The classical fault model for digital circuits and busses, similar to those found in cyber-physical systems, can be summarized as follows:

- *Single Stuck-at Fault Models*: It is one of the first introduced fault models which is common up to now. In this model, faults are represented as a node having a fixed logic value (stuck-at-0 or stuck-at-1). Such a kind of faults is permanent, while the basic functionality of the circuit is not altered. The main advantage of employing this model is its conceptual simplicity. Current research efforts still examine the single stuck-at fault model. However, this model does not provide an accurate representation of the physical defect's behavior [1].
- *Bridging Fault Models*: It models two signals shorted together. Such kind of fault may change the circuit's sequential behavior, while the voltage level at the end of two shorted wires can massively depend on the location of the bridge. A problem with this kind of fault is that they cannot be predicted in advanced. For instance, a strong driver will for sure overpower a smaller transistor if their outputs are shorted but two

equally strong drivers will probably generate an unpredictable value as their common output.

- *Open Fault Models*: Similar to bridging faults, this kind of defect is common in CMOS processes, since latest architectures incorporate more metal layers (and hence much more vias). Additionally, the copper interconnects make reliability issues even more critical [2]. This kind of fault fixes the gate of a transistor at the open value, and hence the transistor cannot be switched on. Predicting the behavior of a circuit with a (resistive) open is a difficult task, which has not yet fully understood [1].

In addition to that, it is possible to categorize the system faults depending on their duration. More specifically, based on the duration of the faults, they can be classified into three categories: (i) transient, (ii) intermittent, and (iii) permanent.

Transient faults are single errors in logic caused by an event external to that logic (e.g., cosmic radiation, contaminated package) [13]. The impact of transient fault to the proper system's functionality is temporary, whereas it is almost very limited the probability for a hardware resource to be affected by two consecutive transient faults. However, since the consequences of transient faults usually cannot be estimated, recalculation is almost assured to result in a correct output. Specifically, whether, or not, a fault results in failure depends on whether and how the error propagates to the system (e.g., shared memory, or output).

The occurrence of transient faults is tightly firmed to the scaling of process technology, since as transistors shrink, less charge (deposited due to the external event) is necessary to trigger a fault [4]. This problem becomes far more important with the continued increase of number of transistors that are integrated to the CPS architectures. Research work was already devoted to the study, detection, and correction of transient faults.

The available design approaches incorporate techniques in order to reduce the likelihood of transient errors to result to enormous output. For instance, transient errors in processing cores are mitigated through the careful instantiation of functional units with variable resilience so that reliability, performance, and cost constraints are met [12]. Similarly, there are also software tools that automatically provide sufficient protection to digital systems against single event upsets. The most widely accepted of these tools is based on appropriately triplicating the system's functionality mapped both on logic and memory resources, which is candidate to be affected by these faults [11]. Regarding the interconnect structures, there are available design approaches that include various recovery mechanisms for handling transient errors in data-paths (e.g., implemented through busses, Network-on-Chip, etc.) [14].

These faults can be represented on multiple levels of abstraction. For instance, on the device level, transient faults are represented as voltage, or current, sources. Device level models considering cosmic particles have already been studied with SPICE simulations [3], that were used to represent alpha particle collected charges. Similarly, at the logic level, transient faults are modeled as bit-flip of the current propagating signal, whereas transient faults in memory/storage elements are represented as changes in logic values.

The second class of upsets affects the intermittent faults. There are two differentiations between these faults and those occurred due to transient upsets. Specifically, the intermittent faults are reproducible, while the conditions causing the fault remain in force [9]. For instance, in case the on-chip temperature is higher than a particular threshold, then it is possible to violate the setup time for a flip-flop along the system's critical path. As long as the system is too hot, the fault exists and the system fails. On the other hand, if the system cools down, it returns to the previous (correct) behavior. There are some works dealing with efficient ways to predict when permanent failure will occur based on the detection of intermittent faults [10]. Device degradation tends to precede total failure, and as a result, online test can be used to monitor gradual changes in performance and use this information to predict when timing failure is imminent.

The last class of upsets affects the permanent faults. In contrast to previous approaches, if such a fault occurs onto a device, it is irreversible because it leads to a physically broken device. The only possible way to overcome these faults and provide a correct operation for the architecture is by replacing the faulty component, or working around it. Permanent faults may be further categorized depending on their occurrence (either at manufacturing time, or later). Since it is not possible to overcome from these faults with conventional design and/or algorithmic approaches, throughout this chapter we will not discuss permanent faults any further.

5.5 Countermeasures for a Fault-Tolerant System

One of the most important challenges in designing efficient cyber-physical system affects the insurance that the system's components (e.g., sensors, computational cores, actuators, communication links, etc.) are available and operational despite the possibility of faults. Toward this direction, methodologies and tools have already proposed that try to derive fault tolerant systems.

Fault tolerant control is concerned with making the controlled system able to maintain control objectives, despite the occurrence of permanent or transient faults. Hence, fault diagnosis represents the main ingredient of a fault-tolerant control system. Diagnosing the faults that occurred in a cyber-physical system permits triggering control mechanisms to keep a plant working sufficiently well until the necessary maintenance may be performed. In practice, this feature results in a significant improvement in cyber-physical system's safety, productivity and time in service.

The majority of fault diagnosis techniques currently in use and each has its own basic support theory. The first class of methodologies used for fault diagnosis-related problems are based on mathematical models of the monitored plant. The differences between the plant model and its actual behavior are called residuals and form the basis for deciding if a fault did or did not occur; and if a fault has occurred, deciding which particular fault occurred. Unfortunately, these techniques provide satisfactory results only either when plants exhibit linear behavior or when the modeling errors can be kept within acceptable limits. Accurate mathematical models can be obtained only for plants with low behavioral complexity.

There are five key elements in a comprehensive approach to fault-tolerant design, as it is depicted in Figure 5.3: namely *avoidance, detection, containment, isolation,* and *recovery.* Ideally, these are implemented in a modular,

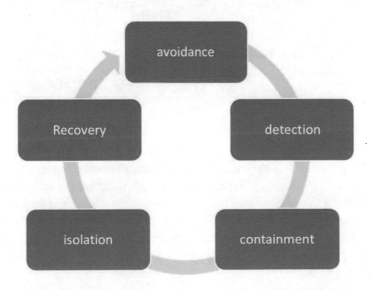

Figure 5.3   Countermeasures for a fault-tolerant system.

hierarchical design, and encompassing an integrated combination of hardware and software techniques. Moreover, the fault-tolerant techniques can be applied at different layers of abstraction in order to balance the desired fault coverage with the performance/cost/design overheads. This approach results in numerous possibilities for fine tuning the performance of the fault tolerance implementation by combining the (sub-)set of fault tolerant elements with the (sub-)set of CPS layers. As an example, error detection may be implemented in the data layer, and recovery may be realized either in the data layer (e.g., if an error correcting code is used), or at the application layer. In a more generic manner, the partitioning and derivation of requirements, and the partitioning and implementation of fault/failure management techniques must be realized in a hierarchical fashion. For each hierarchical level, the existence of appropriate metrics allows the designers to have full control and understanding of the implications that a particular fault tolerant implementation will have on the operation of a CPS subsystem.

As we have already mentioned, these mechanisms are applicable either at hardware, or software level. Even though the hardware-based approaches are much faster compared to the software solutions, the increased fabrication cost, as well as the additional effort required for design modifications, makes the software-based solutions most widely accepted.

For illustrating the set of metrics discussed in this chapter and their usability, we consider a simple example of an application running on a CPS platform consisted of multiple components. The application employs two components of CPS platform $P1$ and $P2$, as it is depicted in Figure 5.4. Apart from the studied mesh-based topology, the concepts introduced throughout this chapter affect also fault tolerant techniques for decision-making in CPS platforms, where individual components are interconnected with any other topology.

5.5.1 Fault Avoidance

The goal of this task is to reduce or eliminate, if possible, the chances for an error to occur. Different mechanisms are possible to be employed toward this goal, while the majority of them are based on some form of redundancy. Unfortunately, hardware redundancy imposes performance degradation in terms of delay, power consumption and area utilization, due to additional infrastructure that imposes. Even though such a limitation might be affordable based on the system's specifications, in advance of applying any hardware-based solution targeting to fault tolerance at CPS platform, designers have

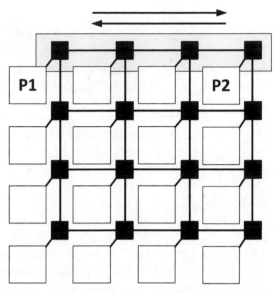

Figure 5.4 A number of nodes connected with mesh-based topology.

to study carefully the impact of their decisions onto the system's QoS requirements.

On the other hand, if there is a demand for realizing fault avoidance at software level, the most appropriate way is by incorporating error detection and error correction codes. These approaches insert to the data either parity information or the error correcting codes. Then, whenever the receiver identifies an error at the transmitted data, this error can be overcome. This is feasible by requesting data re-computation, or by incorporating dedicated hardware targeting to also support error correction functionality.

Usually, fault avoidance techniques can be realized through information redundancy, or hardware redundancy (N-modular redundancy being a typical example). Depending on the targeted number of errors to correct, coding/decoding hardware blocks may require a certain number of cycles to perform their operation. The associated time overhead adds to the total latency of the data being processed and transferred among CPS architecture. We define $T_{av,ov}$ as the time overhead of an avoidance scheme, and compute it as the difference between data latency with ($L_{at,av}$) and without (L_{at}) fault avoidance:

$$T_{av,ov} = L_{at,av} - L_{at} \qquad (5.1)$$

The difference between various implementations of this concept can be significant relative to this metric. Regarding the example in Figure 5.4, if coding/decoding functions are implemented at each architecture's component on the path between $P1$ and $P2$ (component-to-component avoidance), the resulting time overhead will be significantly higher than in the case where only end-to-end avoidance is implemented (i.e., data is encoded at the source and decoded at destination).

5.5.2 Fault Detection

The next hierarchical level in a fault-tolerant design is the detection of faults that were not handled by the avoidance mechanism. Detection is built in most error-correcting codes, which generally can provide information regarding the number of un-corrected faults, when the correction mechanism fails (or is not even present in the particular implementation). Fault detection is then used to assess the need for recovery from potentially uncorrected fatal failures. The quicker the detection mechanism signals the presence of un-corrected faults, the quicker the recovery can be initiated.

Typically, the error detection at CPS platforms is applicable with two alternative approaches: either at component level, or with an end-to-end approach. Specifically, the first approach assumes that error detection is performed at each component of CPS' architecture, in contrast to the latter solution where errors are detected at the end of a path. Even though the second approach (end-to-end check) provides higher fidelity for an error-free decision making, it imposes an increased latency whenever a fault is detected and consequently, the corrupted data has to be re-computed. Hence, the error detection between consecutive switches is usually employed for real designs.

We define the detection latency T_{lat} as the amount of time between the moment a fault occurs and the moment it is detected. Going back to our example in Figure 5.4, fault detection may be performed by the processes $P1$ and $P2$ whenever data is received (end-to-end detection), at the input ports of the intermediate components (component-to-component detection), or at each component input/output port (code-disjoint detection).

5.5.3 Containment

Fault containment is concerned with limiting the impact of a fault to a well-defined region within the CPS. Error containment refers to avoiding the propagation of the consequences of a fault, the error, out of this defined region.

The definition of these regions should be carefully applied in order to ensure that there is no overlap between them. Otherwise, faults that occur at a given region will affect hardware resources assigned to the rest (overlapped) regions.

Fault containment regions (FCR) may be defined with variable resolutions, directly correlated with the quality and resolution of the fault detection mechanism. Regarding the studied usecase depicted at Figure 5.4, and assuming an end-to-end detection mechanism, the fault containment region can be defined as the entire shaded route between the architecture's blocks where components $P1$ and $P2$ are being executed.

It is essential that FCR are independent, in the sense that a fault occurring in a FCR does not affect a different FCR. In this respect, if two paths between components $P1$ and $P2$ can be found that are on independent FCRs, and a fault is detected on one of the decision-making path, the other route can be used to provide an alternative path between the system's components $P1$ and $P2$.

5.5.4 Isolation

The independency of fault containment regions can only be achieved if an effective isolation method can be provided, which can guarantee that the effect of a fault occurring in a FCR does not propagate to another FCR. At the physical layer, in the case of permanent faults, isolation can be accomplished by marking or disconnecting the faulty CPS components (sensor, processing core, actuator, and link) and avoiding their use until, eventually, hardware recovery/repair can be performed through reconfiguration. At higher layers, erroneous data can be dropped on the fly or at the destination process, such that they are not allowed to interfere with the rest of the data and propagate at application level.

5.5.5 Recovery

The ultimate goal of fault-tolerant schemes is to provide means to recover from occurrence of failures. For fault-tolerant and QoS constrained CPS, it is important to recover from failures within the time budget allowed by the QoS specifications. Late recoveries, even when successful, are not acceptable, since they lead to out-of-specification behavior of the CPS subsystem. Consequently, we define recovery time (T_{rec}) as the amount of time that passes between the detection of a fault and recovery from the corresponding failure. A simple form of attempting recovery of erroneous data affecting control of components $P1$ and $P2$ is to provide a hardware correction at lower layers, or a retransmission mechanism at higher layers where, upon detection of an error.

5.6 Improving Fault Masking with Redundancy

Due to the importance of fault tolerant problem, up to now research community has proposed various approaches towards achieving the desired level of fault masking. Common to all these approaches is a certain amount of redundancy. For our purposes, *redundancy* is the provision of functional capabilities that would be unnecessary in a fault-free environment. The concept of incorporating redundancy in order to improve the reliability of a system was pioneered by John von Neumann in the 1950s in his work *"Probabilistic logic and synthesis of reliable organisms from unreliable components"*.

There are two main directions in order to provide fault diagnosis and tolerance systems: either to use *hardware redundancy*, or to use *analytical redundancy*. More precisely, hardware redundancy is based on multiplication of physical devices (multiple instances) and, usually, a voting system to detect the occurrence of a fault and its location in the system. Originally, redundancy techniques were used for coping with the low reliability of basic hardware components. Designers of early computing systems triplicated low-level components such as gates, or flip-flops, and used majority voting to correct faults. As the reliability of basic components improved, redundancy was shifted to higher levels. Larger components, such as memories or processor units, became replicated. This decreased the size and probability of failure of voters relative to that of redundant components. The main drawbacks in this approach is the significant cost for the necessary extra equipment, as well as the performance overheads (e.g., execution run-time, power consumption, etc.). Moreover, the use of redundancy does not immediately guarantee an improvement in the cyber-physical system's dependability, as the overall increase in complexity caused by redundancy can be quite severe. It may diminish the dependability improvement, unless redundant resources are allocated in a proper way. A careful analysis has to be performed to show that a more dependable system is obtained at the end. On contrary, analytical redundancy uses redundant functional relationships between variables of the system. The majority of software-based fault detection and fault tolerant techniques, such as additional check bit attached to a string of digital data, or a few lines of program code verifying the correctness of the program's results, rely on analytical redundancy. The main advantage of this approach compared to hardware redundancy is that no extra equipment is necessary. On the other hand, since fault detection and masking is performed in software level, the penalty in term of execution run-time is higher compared to the corresponding hardware implementations.

Another classification of the alternative fault-tolerant techniques is based on the way that redundancy is applied. Two alternative approaches are commonly used for this purpose, namely the *space redundancy* and *time redundancy*. More specifically, space redundancy is the commonly employed redundancy scheme, where multiple instances of system's components, functions, or data items that are necessary for fault-free operation are replicated. Space redundancy is further classified into hardware, software, and information redundancy, depending on the type of redundant resources added to the system. On the other hand, in time redundancy the computation or data transmission is repeated (usually with a delay) and the result is compared to a stored copy of the previous result.

Finally, the redundancy scheme is applicable in *passive*, *active*, or *hybrid* manner. In case the passive redundancy approach is selected, the emphasis is at the fault masking rather than their detection. This technique insures that the system's output data is fault-free in spite of the presence of faults. Typical application domain of passive redundancy techniques is the high-reliability applications, where even short interruptions of system operation are unacceptable, or in usecases where it is not feasible to repair the system (e.g., deep-space applications). Representative implementation of this technique is the N-modular redundancy, and more specifically the widely employed triple modular redundancy scheme. In contrast, active redundancy performs fault masking by first detecting the faults which occur, and then performing the actions of location, containment and recovery are performed to remove the faulty component from the system. Typical domains where active redundancy might be applied include among others applications requiring high availability, such as time-shared computing systems or transaction processing systems, where temporary erroneous results are preferable to the high degree of redundancy required for fault masking. Although it is not a pre-request, usually in order to overcome the fault operation, system has to be stopped and reconfigured. Finally, the hybrid redundancy is a mixture of the previously two approaches. Similar to passive redundancy, the current one is used to prevent generation of erroneous results by providing mechanisms that perform fault detection, location, and recovery. In contrast to active redundancy, the hybrid approach enables reconfiguration with no system downtime.

5.7 Fault Forecasting

Due to the importance of faults in cyber-physical systems, especially for those that target mission-critical application domains, various methodologies and techniques have been proposed aiming to estimate how faults are present in the system, how these faults are propagated among system's components, possible future occurrences of faults, as well as their consequences. These approaches are also known as fault forecasting. Their functionality relies on evaluating the system's behavior with respect to fault occurrences or activation either with qualitative or quantitative way. More precisely, the qualitative way aims to rank the failure modes or event combinations that lead to system failure. On contrast, the quantitative approach focuses to evaluate in terms of probabilities the extent to which some attributes of dependability are satisfied.

5.8 Conclusion

A number of mechanisms that are able to overcome limitations posed by the existence of upsets, or faults, during the operation period were explored. These techniques are applicable either at hardware or software level. Additionally, they can be performed at different levels of abstractions in order to guarantee the desirable level of fault masking. Although most of them impose some form of "performance overheads (e.g., in terms of operation frequency and power/energy consumption), they are usually assumed as pre-request due to the criticality of target application domain.

References

[1] Chow, E. Y., and Willsky, A. S. (1980) "Issues in the development of a general design algorithm for reliable failure detection," in *1980 19th IEEE Conference on Decision and Control including the Symposium on Adaptive Processes* (IEEE: New york, NY), 1006–1012.

[2] Chen, J., and Patton, R. J. (1999). *Robust model-based Fault diagnosis for dynamic systems*. Norwell, MA: Kluwer Academic Publishers.

[3] Srinivasan, J., Adve, S. V., Bose, P., and Rivers, J. A. (2005). Lifetime reliability: toward an architectural solution. *IEEE Micro*, 25, 70–80.

[4] ITRS. (2011) Itrs roadmap. Available at: http://www.itrs.net, 2011. [accessed April 30 2017].

[5] Abraham, J. A., Krishnamachary, A., and Tupuri, R. S. (2002). "A comprehensive fault model for deep submicron digital circuits," in *Proceedings of The First IEEE International Workshop on Electronic Design, Test and Applications* (IEEE: New york, NY) 360–364, 2002.

[6] Aitken, R. C. (1999). Nanometer technology effects on fault models for IC testing. *Computer*, 32, 46–51.

[7] Baumann, R. (2005). Soft errors in advanced computer systems. *IEEE Design Test Comput.* 22, 258–266.

[8] Borkar, S. (2005). Designing reliable systems from unreliable components: the challenges of transistor variability and degradation. *IEEE Micro.* 25, 10–16.

[9] Tosun, S., Mansouri, N., Arvas, E., Kandemir, M., and Xie. Y., (2005). "Reliability-centric high-level synthesis," in *Proceedings of the Conference on Design, Automation and Test in Europe, Vol. 2*, DATE '05 (IEEE Computer Society: Washington, DC), 1258–1263..

[10] Xilinx. (2017). Xilinx tmr tool. Available at: http://www.xilinx.com/ise/optional_prod/tmrtool.htm [accessed April 30, 2017].

[11] Murali, S., Theocharides, T., Vijaykrishnan, N., Irwin, M. J., Benini, L., and De Micheli, G. (2005). Analysis of error recovery schemes for networks on chips. *IEEE Design Test Comput.* 22, 434–442.

[12] Maheshwari, A., Burleson, W., and Tessier, R. (2004). Trading off transient fault tolerance and power consumption in deep submicron (dsm) VLSI circuits. *IEEE Trans. Very Large Scale Integrat. Syst* 12, 299–311.

[13] Constantinescu, C. (2003). Trends and challenges in vlsi circuit reliability. *IEEE Micro* 23, 14–19.

[14] Smolens, J., Gold, B., Hoe, J., Falsafi, B., and Mai, K. (2007). "Detecting Emerging Wearout Faults," in *Silicon Errors in Logic – System Effects Workshop*.

6

A Framework for Research and Prototyping in Robotics: From Ideas to Software and Hardware Development

Konstantinos Machairas, Spyridon Garyfallidis, Iosif S. Paraskevas and Evangelos Papadopoulos

Department of Mechanical Engineering, National Technical University of Athens, 15780 Athens, Greece

Abstract

Development and prototyping of robotic systems requires the involvement of many people and many hours of design, development, and cooperation; significant time and effort overhead is required for evaluating conceptual ideas in design, control and technology, and for bringing them fast into reality for testing. Based on the important advances of the last decade in hardware and software, a simple and low-cost framework and its underlying ideas are presented, with steps that aim at accelerating robotics research work in academia and industry. The framework's functionality is validated and illustrated by two application examples concerning the control systems of a single-legged hopping robot and an instrumented treadmill. The software required to conduct the same experiments is provided, with the intention to help the reader reuse it in similar applications.

6.1 Introduction

Cyber Physical Systems (CPSs) are electromechanical and mechatronic systems interacting through physical actions with their environment (e.g., motorized motion), controlled by interconnected embedded computers. The field of CPSs builds on embedded systems theory with a great focus on how subsystems communicate to form complex systems with more capabilities. CPS research aims at a deep understanding of small embedded components, and at their best possible integration into a whole system with increased functionality [1]. The fields of dynamics, mechanical design, electrical/electronic design, software design, control, and networking, assume equally important roles and compose the multidisciplinary nature of the field of CPSs [2]. Apparent is the relation to robotic systems theory, which strongly builds on advances in these same fields, and also to the recent developments of the Industrial Internet of Things (IIoT) [3] and Industry 4.0 [4], and their promise for smarter industrial procedures. During the last decades, CPSs have gained attention as an important domain, bridging multiple theoretical and applied technological fields toward promising application goals ranging from smart devices to smart cities, and aiming to have strong impacts on most sectors of human life (e.g., health, transportation, etc.) [5]. This chapter focuses on CPS design, control, simulation and development.

In regard to the recent advances in robotics, an increasing tendency of the community is witnessed to provide low cost and easy to implement solutions concerning the theoretical analysis of a system (modeling, design, simulation, and control), as well as its development (embedded hardware and software). Already, important advances have led to big strides in software (e.g., the ubiquitous use of the Robotics Operating System or ROS [6]), and in hardware (e.g., the constantly increasing number of embedded single board computers with high computational power and low cost [7]). However, the robotics community still lacks a complete framework involving all the necessary stages that lead from ideas to prototype development and testing. There is a lack of standard procedures to help research teams create and share design and development tools and build upon existing work to minimize time and maximize research quality. The challenge here is to connect the pieces from diverse science and technology fields toward coordinated steps in the field of CPSs.

In this chapter, an attempt is made to provide new tools and also combine existing ones in software and hardware with the goal to create a complete framework to aid researchers bring conceptual ideas into reality in minimum time and at minimum cost. To this end, the focus is on widely adopted tools and techniques. Also all necessary material for anyone to reuse it in a similar or modified manner is provided. First, the modeling and simulation stages are addressed for robots with multiple degrees of freedom (dof) subject to frequent collisions with the environment – a typical class of problems – and template simulation software is provided. Second, simple steps toward the design and development of prototypes are proposed, based on the idea of easily interconnected, and low cost embedded computers. Finally, two application examples – mostly concerning motion control – are presented to exhibit the simulation and prototyping aspects of the proposed framework.

6.2 A Framework for Simulation and Prototyping

The proposed simulation and prototyping framework provides solutions for the stages of modeling, simulation, control, and system development. In this section, each stage of the procedure is properly described, and simple software and hardware solutions are provided for a class of problems commonly addressed in robotics, namely motion control and interaction with the environment.

The first stage includes finding the most important unsolved problems and proposing concepts and ideas toward their solutions. In this scope, this stage mostly concerns ideas in the areas of design, control and technology. Next, simulation experiments must be conducted to provide the first proof of concept, hardware prototypes must be designed and built, and finally hardware experiments must validate the previously derived theoretical results.

6.2.1 Modeling, Dynamics, and Simulation

6.2.1.1 Dynamics derivation and simulation methods

Before conducting hardware experiments, simulation experiments are conducted to first evaluate the theoretical idea or concept. To this end, a model of the system must be built, and then simulated. Two options are typically followed and both present their own advantages; deriving the respective system of differential equations and then solving it with numerical analysis software, such as Mathematica and Matlab, or use 3D simulation software packages such as Gazebo and Adams [8], for which a physical description with mass properties is only required.

In the first option, the models are kept simple but representative, to aid the dynamics derivation phase, to allow the understanding of each component's role in the equations, and also to be used in the design of model-based controllers. This approach provides a simulation environment, but also tools for design and control. The difficulties that arise concern the derivation of the often intricate and multi-dimensional system of equations, the tedious programming, and the art of selecting the most important features to model.

Template Mathematica and Matlab files for deriving and solving the Equations of Motion (EoM) for a model of the quadruped robot Laelaps (Figure 6.1) – designed and built in the Control Systems Laboratory (CSL) of the Mechanical Engineering Department of NTUA – are provided as supplementary material for this chapter [9]. The EoM are derived in symbolic form using Mathematica, and then imported for integration and animation in Matlab; this approach was followed in designing and simulating the trotting controller proposed in [10]. The software is designed such as to be easily reconfigured for application in other systems of the same class, namely multibody dynamical systems subject to impacts with the environment. The reader is encouraged to employ the code as a tool for dynamics derivation and simulation in related applications.

In the second option, in which a 3D dynamics simulation software package is used, a detailed model can be built easily. The focus here lies on tuning the simulations so as to eliminate, if possible, the expected differences from the experiments. A comparison of the best known and most used 3D simulation packages for use in robotics can be found in [8], where Gazebo appears to be a prevalent solution in terms of cost, community support and ability for interconnection with other software like ROS, ubiquitously used for robot control [6]. This exact combination – Gazebo for modeling and simulation, and ROS for control – is proposed here for 3D simulations. Application examples are presented next to show the packages' ease of use, functionality, and compatibility with the real robot's firmware.

(a)

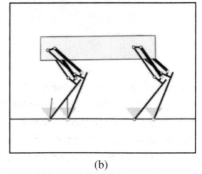
(b)

Figure 6.1   (a) The quadruped robot Laelaps, designed and built at CSL. (b) The 2D model of Laelaps animated in the simulation environment built in Matlab.

6.2.1.2 Modeling the environment

Depending on the application, one has to model the environment with the proper level of detail. In robotics, this includes, among others, the modeling of walls, gravitational acceleration, terrain, aerodynamic or hydrodynamic forces, impacts, etc. The descriptions are such that follow simple or complex forms of physics equations: for example it may be enough to describe a wall as a static object; then, a mobile robot has to avoid certain values in the location of the object in order to avoid any interaction with the object. On the other hand, it is common to require a description of interactions, because the robot operates in a way that these interactions are unavoidable; the foot-terrain interaction is a perfect example which will be analyzed a bit further to give an insight on how to work on these cases.

The interest in this example is how to model impacts between a legged robot and terrain, using a method which shall describe the interaction between different materials, compliant (i.e., able to have deformations) and non-compliant, retaining a high level of fidelity. The answer to this question is important, because a well-established model is necessary for the accurate representation of impacts on simulations. In principle, the impacts can be modeled via three methods: the stereomechanical theory method, the Finite Element Method (FEM), and the compliant/viscoelastic approach [11]. Each method has its pros and cons but the use of the viscoelastic method seems more appropriate, as the impact between different materials can be described by lumped parameter models with suitable characteristics. There are various viscoelastic models in the literature with more prominent the Hunt–Crossley (HC) model, [12]; in fact the majority of the viscoelastic models use the HC model as a basis and this will be also the basis here. In HC, the interaction force is calculated by,

$$F_g(y_g, \dot{y}_g) = k_g \cdot y_g^n + b_g \cdot \dot{y}_g \cdot y_g^n \qquad (6.1)$$

where k_g and b_g are the stiffness and damping coefficients respectively, n in the case of Hertzian non-adhesive contact is equal to 1.5, and y_g is the depth of interpenetration. In Figure 6.2(a) the shape of a typical HC impact is given. The area inside the curve is the non-recoverable energy dissipated during impact inside the materials due to mechanisms like internal vibrations and local plastic deformations.

However, the behavior of real materials is somehow different according to the experimental results in the literature. In Figure 6.2(b) the experimental results of the impact of a metallic sphere on a rigidly supported thin laminate

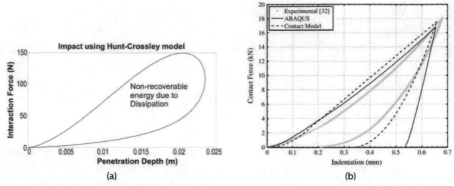

Figure 6.2   (a) Typical HC interaction force–penetration depth diagram. (b) Force–indentation response of a metallic sphere impacting a rigidly supported thin laminate.

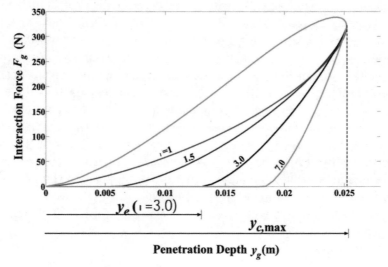

Figure 6.3   Impact curves for the proposed impact model (2) for various λ.

are shown. The qualitative similarity of the HC model is apparent; however, the HC model fails to predict the permanent deformation analytically [13].

In order to tackle the issues that other impact descriptions have, the authors have proposed a novel impact model which has viscoplastic characteristics. This model shows very good correlation with experimental results found in the literature and it can efficiently describe a large number of interactions that occur in robotics (in terrestrial applications and also in space). Briefly, the model calculates the interaction force F_g at impact instance i by,

$$F_g^i(y_g, \dot{y}_g) = \begin{cases} F_c^i = \left(\lambda_c^i k_g + b_g \dot{y}_g\right)\left(y_g - y_e^{i-1}\right)^n, & \dot{y}_g \geq 0 \\ F_r^i = \left(\lambda_r^i k_g + b_g \dot{y}_g\right)\left(y_g - y_e^i\right)^n, & \dot{y}_g < 0 \end{cases} \tag{6.2}$$

where subscript c stands for compression, r for restitution, y_e is the penetration depth, and the index i identifies the impact instance, while the term *Coefficient of Permanent Terrain Deformation* λ is proposed. The result of the model can be seen in Figure 6.3, however the interested reader should consult [14] for details.

6.2.2 System Development: Hardware and Software

6.2.2.1 Introduction

In this section, methods for fast prototyping of robotic systems of various structures and complexity levels are presented, which are scalable and can be implemented relatively easy. Since hardware experiments provide the final and most convincing proof of the claims made in the analysis and simulation phases, it is critical to be able to design, build, and test a prototype in a fast, low cost, and debuggable way. The focus here is on hardware and software development for robotic applications like legged robots, which include modules with low level read and write capabilities, e.g., microcontroller units (MCUs), controlled through a network. Such systems, also known as *Networked Control Systems* (NCSs) [15], are inextricably linked to CPSs and will be the main subject of the following paragraphs. An example of an NCS is shown in Figure 6.4, where sensor measurements that are used as feedback, and control signals that are used as controlled reactions, are exchanged via the network.

NCSs consist of distributed nodes and are typically controlled by one or more *Single Board Computers* (SBCs) running an operating system, or even a higher level meta-operating system like ROS. The number of nodes

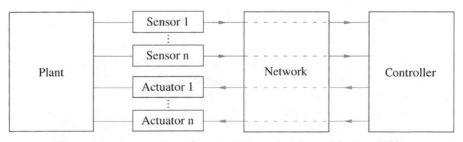

Figure 6.4 The structure of a typical Networked Control System (NCS).

comprising an NCS is indicative of the system's complexity. Common applications for a node include controlling a DC motor, interfacing with sensors like Inertial Measurement Units (IMUs), LIDAR, or force sensors, and more. These low level devices come with specifications for communication, with common interfaces including Quadrature Encoder Interfaces (QEIs), Pulse Width Modulation (PWM) modules, serial interfaces such as Serial Peripheral Interfaces (SPI), etc. Thus, in the design phase of an NCS, suitable embedded computers must be selected such as to first satisfy these low level requirements. Note that some I/O capabilities like QEIs, or Digital to Analog Converters (DAC) are only found in application specific MCUs, whereas others like SPI are ubiquitous. Besides communicating with the devices, the embedded computers must be able to connect to the network, through which they will communicate with the other nodes. This network can have thousands of nodes in the case of a factory or can be much smaller in the case of a small mobile robot. Therefore, depending on the network type, another requirement is added to the process of selecting the appropriate MCUs for each case. For instance, if it is for a node to be connected to a CAN [16] or an EtherCAT network [17], it must have the respective specifications.

Networking of embedded devices has been a subject of meticulous studies for many years, with industry showing early the way by introducing numerous communication protocols for use in factories, vehicles, medical devices, etc. [18]. Interestingly, there is not yet a universally adopted protocol for industrial communications, as there is, e.g., for the Internet. As networks and related technology fields advance, many new communication technologies emerge, many old ones become obsolete, and the future goals get higher in terms of speed, safety and determinism.

The following paragraphs gradually introduce and discuss the fundamental concepts behind industrial communications, aiming at setting the basis on which the proposed methods will be presented. The main focus is on the aspects of communication speed and determinism, which are critical for real-time closed loop applications. Lastly, the hardware and software aspects of the proposed NCS are described.

6.2.2.2 The automation pyramid

The topics discussed are strongly connected with the old *automation pyramid* idea [19], which separates an industrial network in different logical and topological hierarchical layers. The network types in the pyramid differ in size, complexity, requirements and purpose. On the top, the *Global Area Networks* (GANs) are met, able to cover even intercontinental distances, next are the *Wide Area Networks* (WANs) followed by the *Local Area Networks* (LANs), and below are the *Field Area Networks* (FANs), (Figure 6.5). During the last decades, important advances occurred in both LANs and FANs, aiming at serving different communication purposes; high bandwidth for large data packets delivery for the former, compared to real-time delivery of small data packets for the latter. However, this difference gradually lost importance, as Ethernet started occupying an increasingly wider range in the pyramid, penetrating every level and replacing almost all of the older variants.

Here the interest lies in the lowest levels of the automation hierarchy, where high accuracy embedded control takes place. There, depending on the number of nodes connected to the network, complexity rises and the need of systematic approaches concerning the communication system emerges. Regarding communication at this so called field level, the seminal works on *Fieldbus* technologies, which have been advancing for the last 30 years, have already set a rigid basis under the idea of connecting multiple devices via a bus-like shared medium [20]. This *Fieldbus* revolution pushed industrial communications to a high level, with applications spreading across all technology domains under various standards. Interestingly though, a second promising revolution has recently begun, building upon the widespread *Ethernet* technology [21]; *Ethernet*-based control systems will be rigorously studied herein.

6.2.2.3 The OSI model and Media Access Control (MAC) methods

Regarding the structure of an industrial network, studies refer to the physical and the logical topology as the two key characteristics [22]. The physical topology describes the installation of the nodes (in a star, ring or other formation), the cabling and the hardware required, while the logical topology – not necessarily similar to the physical one – defines the protocol used by the signals to transfer data. Of great importance in networks history was the establishment of the *Open System Interconnection* (OSI) model, as a seven-layer open reference model describing the required hardware and software components for connecting all the – incompatible in the past – types of networks [22]. A clarification needed is that the numerous standards occupying the different layers are not considered parts of the OSI model. According to the model, in a typical communication channel, the transmitting side encodes each data packet following the steps from layer 7 down to layer 1 before sending the package to the receiving side, where the inverse process

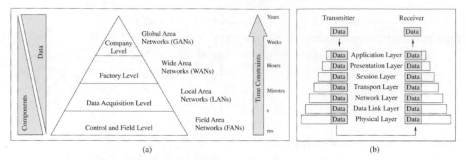

Figure 6.5   (a) The automation pyramid and (b) the OSI model.

takes place. Each layer, in the transmitter prepares data to be read by the peer layer on the receiving side, and then transmits the data to the lower layer to do the same until the packet is finally sent out from the physical layer (Figure 6.5(b)).

The *Media Access Control* (MAC) methods occupy a part of the data link layer of the OSI model, as mechanisms to access the network and manage the bandwidth for achieving the desired communication characteristics. The MAC sub layer has a major role in defining the determinism of the communication; depending on the application, the method may favor large data packet transfer with high and unbounded delays, or small data packets with low and bounded delays. Known MAC methods include *Frequency Division Multiple Access* (FDMA), *Code Division Multiple Access* (CDMA), *Space Division Multiple Access* (SDMA), and *Time Division Multiple Access* (TDMA) [20]. The applications studied here typically use the latter, with which the nodes use the medium sequentially. Referring to TDMA methods, the available sub-strategies include *Polling*, *Token Passing*, *Time-Slot-Based*, and *Random Access* methods [20]. Ethernet adopts the *Random Access* strategy, and thus it is of our main interest in this chapter.

Random Access methods include various *Carrier Sense Multiple Access* (CSMA) methods; some are modified for collision avoidance (CSMA/CA), as for example in CAN, and others for collision detection (CSMA/CD), as in Ethernet's original version with shared medium [23]. Focusing on CSMA/CD, collisions are immediately detected by the sending nodes, which monitor the bus while sending. After collision detection, the nodes abort the data transfer and wait for a random time before trying again, clearly showing a nondeterministic behavior, unsuitable for control applications. However, with the introduction of switched *full-duplex* Ethernet, collisions can now be avoided as described next.

6.2.2.4 Networked Control System design

When designing an NCS, there is no rule of thumb that one should always follow. Taking into account the available software and hardware and their costs, merits and demerits come to the surface and point toward the final solution. In this chapter, we propose a simple but powerful control network structure based on Ethernet and the *User Datagram Protocol* (UDP) [22]. The Ethernet standard is the most common way to connect devices in a local area network, and it occupies layers 1 and 2 of the OSI model. So far, Ethernet was considered mostly for the upper levels of the automation pyramid, but recently it has gained ground toward the lower field levels with industrial standards ensuring high-speed, reliability, and determinism [23]. Moreover, besides the usually costly industrial solutions, simple and cost-effective nonindustrial UDP implementations have been also shown for real-time embedded systems [24–26].

In general, Ethernet technology presents many advantages for use in CPSs; here, we sum up some critical aspects. Firstly, it is a fast, inexpensive technology that can suit many purposes. It can transfer data ranging from short messages to big files and over long distances, it has minimum requirements in hardware and software design – since related hardware modules and software packages are widely available – and it can take advantage of numerous higher-level protocols such as IP and UDP, or other industrial variants like EtherCAT [27]. Finally, most computers have Ethernet support built in, which favors the direct interface of an Ethernet-based CPS with a personal computer.

Throughout the next paragraphs, a generic case is analyzed to show how an NCS can be built using the herein discussed toolsets. Consider a common application, where n DC motors are connected to n MCUs and are controlled through a switched Ethernet network by a computer running ROS, Figure 6.6. To analyze the proposed NCS, the important aspects of Ethernet and UDP in terms of determinism and system performance are examined.

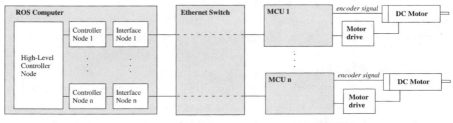

Figure 6.6 The architecture of the proposed NCS.

Also, software design methods for the MCU nodes and the ROS nodes are rigorously discussed.

6.2.2.5 Switched Ethernet and determinism

Before switches, Ethernet networks used hubs. In a hub-based network, packets received by the hub were broadcasted to all ports, raising the risk of message collisions, and packet retransmissions after randomly long waiting times, as defined by CSMA/CD. On the other hand, a switch is a device that can learn the network's topology, divide it into different collision domains, and send packets only to their destination ports. The main switch types are *store and forward*, which examine the whole packet before transmitting, and *cut through*, which immediately forward the packet to its destination port. Store and forward is slower, but it is the most common type of switching.

Regarding sending and receiving packets through the same wire, another risk for collisions emerges even for switched Ethernet. In the early *half-duplex* Ethernet, where a node could not send and receive data at the same time, emerging collisions were addressed with CSMA/CD resulting again in a nondeterministic communication type. However, when *full-duplex* Ethernet was standardized, every node was provided with a unique collision domain allowing for sending and receiving simultaneously, while doubling the available bandwidth [28].

Conclusively, switched Ethernet can assume a deterministic nature, since collisions can be prevented, and thus no random waiting times are required. Yet, a last point requires the designer's attention to ensure the desired real-time character. In case a switch port receives a large number of packets, the port's buffer may overflow and lead to unexpected delays or data loss [29]. By default, switches use first-in-first-out (FIFO) queues. That said, to avoid the risk of data loss and to ensure a high level of determinism (i.e., bounded latency), a control network should be designed such that it would not need large buffers in nominal operation. To this end, congested segments must be avoided early in the design phase. Depending on the hardware components used and the network architecture, the proper data rate for each communication path must be defined in software. For instance, say the network is badly designed, and network traffic for the nominal case is intense at several nodes finally leading to always full buffers. Apart from packet dropping, there is also unwanted latency introduced to the system, equal to the time required to empty the respective buffers. Along these lines, the software developer must carefully take into account all the limitations regarding hardware data rates to achieve optimal performance.

6.2.2.6 Quality of Service (QoS)

The QoS parameters refer to measures of how well a network performs according to criteria of timeliness, reliability and other [30]. The bottlenecks concerning these parameters must be identified and taken under consideration in the design phase. To this end, the most important QoS parameters are shortly presented here.

- **Bandwidth**: a measure of the maximum amount of data bits that can pass in a given interval between two network nodes – measured in bps (bits per second).
- **Throughput**: the actual rate that the data are transferred.
- **Delay**: a measure of how long it takes for a unit of data to be transferred from the source node to the receiver node.
- **Jitter**: the variation in packet delay.

6.2.2.7 Latency in switched Ethernet

Adding to the above, **determinism** can be considered a QoS parameter of great significance when referring to NCSs. It refers to the unbounded delays that may occur in any layer of the OSI model. An analysis on the delays introduced in the various stages of Ethernet communication is useful here. The total delay for a communication channel is the main indicator for determinism and can be calculated as follows based on the analyses conducted in [23] and [29]:

$$T_{\text{delay}} = T_{\text{pre}} + T_{\text{wait}} + T_{\text{frame}} + T_{\text{prop}} + T_{\text{switch}} + T_{\text{post}} \qquad (6.3)$$

with the time components defined as follows.

Preprocessing and Postprocessing times: T_{pre} is the time required for the sender to encode the data for sending over the network, and T_{post} is the time required for the receiver to decode the received data. These depend on the devices and they can be the prevalent cause of delay.

Waiting time: T_{wait} is the waiting before transmitting time in case the network medium is unavailable. It depends on the MAC strategy followed and the network traffic.

Frame time: T_{frame} is the time required to send a packet, and can be calculated as:

$$T_{\text{frame}} = (N_{\text{data}} + N_{\text{ovhd}} + N_{\text{pad}})8T_{\text{bit}} \qquad (6.4)$$

where N_{data} is the data size in bytes, N_{ovhd} is the number of bytes used as overhead, N_{pad} the number of bytes required to reach the minimum frame size, and T_{bit} is the time to transmit 1 bit, calculated as the inverse of the data rate (e.g., $T_{bit} = 10$ ns for 100 Mb/s Ethernet).

Propagation time: T_{prop} is the time required to propagate a message between two devices. It is defined by the velocity factor (VF) of the transmission medium, which is the ratio of the speed of a wavefront passing through the medium, to the speed of light in a vacuum [31]. Depending on the insulating material, typical VF values lie between 0.4 and 0.7, with 0.7 corresponding to approximately 2.1×10^8 m/s. T_{prop} is less than 1 μs for distances below 100 m for most networks.

Switch delay: is the time a frame is delayed at the switch, and can be calculated as $T_{switch} = T_{mux} + T_{queue}$, where T_{mux} is the multiplexing delay, after which the switch starts to transmit the frame once it is received, and T_{queue} is the time the frame waits in the switch queue plus the time required to transmit it [29].

6.2.2.8 Message exchange using Ethernet, IP, and UDP

A computer sending a UDP message, first places the message in a UDP datagram consisting of a UDP header followed by the data payload, and then places the datagram in the data area of an IP datagram. The IP address does not contain information about the physical location of the destination, and therefore the datagram is placed in an Ethernet frame that contains this kind of information. Finally, an Ethernet driver sends the packet on the network. Conversely, on the receiving side, the Ethernet layer passes the IP datagram to the IP layer, which removes the IP header and passes the data included in the UDP datagram to the port specified in the datagram's header – Figure 6.7 for a visualization of the procedure.

At this point, a detailed description of the Ethernet frame and the resulting message is needed to better understand the basics of Ethernet communication. Referring to Figure 6.7, the *Preamble* field (PRE) consists of seven bytes of the form 10101010, and is used for bit synchronization. The *Start Frame Delimiter* (SFD) is the byte 10101011, which indicates the start of a frame. The *source* and *destination* MAC *(Media Access Control)* addresses (SA and DA) are the physical hardware addresses consisting of 6 bytes each. The *Ethertype* indicates the sort of data contained in the frame; for an IP datagram, the field would contain the value 0x0800. The *Data* field contains a message of size ranging from 46 to 1500 bytes, including information for IP, UDP or other. Note that if the data size is less than 46 bytes, the remaining bits are padded as zeros. Finally, the *Frame Check Sequence* (FCS) is used to detect errors in a received frame, and the *Interpacket gap* (IPG) contains 12 bytes

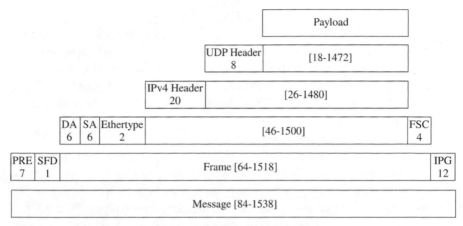

Figure 6.7 The UDP/IP stack.

to cause the required pause between the network frames. Based on these, the size of an Ethernet message always lies between 84 and 1538 bytes.

Details for the path of the data through the *Network*, the *Transport* and the *Application* layers of the proposed architecture are given next.

6.2.2.9 Network layer: The Internet Protocol

The *Internet Protocol* (IP) helps the data packet find the way to its destination on the Internet. Several communications also use IP in local networks to employ its companion protocols, TCP and UDP. It is a connectionless and unreliable protocol, since it doesn't provide flow control and error checking of the payload. The structure of an IP datagram and its header is well described in Kurose and Keith [22]. A protocol field is included in the datagram so the IP layer will know where to pass the received data, Figure 6.7. For instance, for UDP, decimal 17 is used.

6.2.2.10 Transport layer: The User Datagram Protocol

Both UDP (*User Datagram Protocol*) and TCP (*Transmission Control Protocol*) communications are established between logical endpoints called sockets and existing only in software. A socket is defined by a port number and an IP address. The port numbers identify the sending and receiving processes running on the communicating devices. Although in an Ethernet frame, the address fields already identify the communicating interfaces, UDP and TCP precisely specify the source and destination nodes by naming the respective ports. Here we focus on UDP, which only adds port addressing and optional error detection to the message being sent. Unlike TCP, it is a connectionless protocol, meaning that a computer can send a message without establishing if the target computer is available on the network. These make it unreliable but simpler to implement and faster. UDP can also send a message to multiple destinations at once, while TCP cannot. Those said, and based on the considerations made about the protocol's determinism and reliability, it is considered more suitable than TCP for the discussed control applications.

6.2.2.11 Application layer

The data payload contained in the UDP datagram (Figure 6.7) must also follow a protocol to let the receiving application know what to do with the incoming data. An application may use a standard protocol such as the hypertext transfer protocol (HTTP) for requesting and sending Web pages, the file transfer protocol (FTP) for transferring files, or the simple mail transfer protocol (SMTP) for exchanging e-mail messages. Other applications however may use simpler custom protocols, e.g. in an embedded system like the one shown in Figure 6.6, an application may periodically receive sensor measurements and use the data to control motors, relays, or other circuits.

Before further discussing the proposed software structure at the application layer, putting the possible communication types into categories will help define the nature of the proposed NCS and understand the logic behind the design. A first categorization can be made into *cyclic* and *acyclic* communication types. Connectionless services – like the ones used here – are typically used for cyclic data exchange in the sense that lost packages are not being sent again delayed and outdated, favoring new data to be sent. In practice, buffers are used in a FIFO logic at all nodes to store the most recent data. On the contrary, acyclic communications typically use protocols with reception acknowledgement and packet retransmission mechanisms. A second categorization would be into *time-triggered* and *event-triggered* communication types, with the former mostly used for periodic data exchange, as in the presented case. Finally, regarding the way the central control node communicates with the distributed slave nodes, the communication types can be based on the *client–server* or the *publisher-subscriber* paradigms [20]. For instance, the processes (nodes) in a ROS system typically communicate using a publisher-subscriber type of communication.

6.2.2.12 Software design for the MCU node

On the MCU side, the software must handle the signal level communications, and also set up the node's connection to the network. Specifically in the motor control case, the MCU must read the signals from the encoder attached to the motor, and send a PWM signal to the motor drive. Here, instead of closing the loop on the MCU, it is preferred that the loop is closed on the ROS computer in order to use its rich toolsets for control. To this end, the MCU must send the encoder measurements via UDP to the ROS computer to calculate the control output and send it back to the MCU also via UDP.

Along these lines, a suitable MCU is selected such as to have motor control features and Internet connection capabilities. Interestingly, a very limited number of ready-to test and low cost development boards with such specifications are available in the market. The Tiva^{TM} C Series TM4C1294 Connected LaunchPad [32] by Texas Instruments is a good candidate, since it features an 120MHz 32-bit ARM Cortex-M4 CPU, Ethernet connectivity and a QEI module, at the price of $20. Also provided by Texas Instruments is the TivaWare library, which significantly accelerates the software development. Those said and without loss of generality in the design process, TM4C1294 is used in the applications presented herein.

In the n-motor control case shown in Figure 6.6, n TM4C1294 boards are used. Encoder measurements are sampled from the boards and sent with a desired frequency to the ROS computer via UDP to calculate and send back the corresponding control outputs. The choice of the sampling frequency is of great importance to achieve the desired system performance; a too slow sampling rate will give low resolution and a slower control loop, while a too fast sampling rate will lead to a saturated network and data loss [23]. In terms of software design, a timer is used to strictly define the sampling and the transmission rate.

In general, the achievable data rate is limited by several factors; an important one is the speed of the connection. Typical Ethernet speeds are 10 Mb/s, 100 Mb/s, and 1,000 Mb/s. For a full-duplex 100 Mb/s connection and a minimum message size of 84 bytes (Figure 6.7) the minimum frame delay T_{frame} is $(84 \times 8)/(100 \times 10^6) = 6.72\,\mu s$, corresponding to a maximum frequency of 148.81 kHz. This delay is fixed if the useful data packet size is less than the minimum size of 18 bytes. For a full message of 1538 bytes the theoretical minimum delay is $(1538 \times 8)/(100 \times 10^6) = 123.04\,\mu s$, corresponding to a maximum frequency of 8.1274 kHz. In control applications the messages are short and so the frame delays are expected to be close to 10 μs, based on the previous calculations. Interestingly, there is no difference in delay when sending 1 and 18 bytes of useful data; the optimal exploitation of the frame would be to send 18 bytes of information.

Although, the above discussed network delay can become a bottleneck as the system size increases and the performance requirements raise, the node processing delays are those that typically dominate in small networks. The processing power of the hardware, and the software design determine

Table 6.1   The structure of the message contained in the UDP data frame

Message ID	Data	Description
0x31	1 byte (PWM duty cycle)	PWM and DIR command
0x42	4 byte (encoder)	Encoder measurement

how fast the data can be processed at the nodes. Referring to (3), besides the already discussed T_{frame}, the components T_{pre}, T_{post} and T_{switch} are the most important for estimating the total delay, since the waiting time T_{wait} and the propagation time T_{prop} are negligibly small. To achieve the desired T_{pre} and T_{post}, suitable embedded computers must be properly selected and programmed. Finally, to keep T_{switch} to a minimum the data rates must be set in software such that segment congestions are always avoided.

The calculations made set the first limitation for the maximum achievable frequency at 150 kHz. Adding also the node and the switch delays this frequency is further reduced. However, since it is hard to estimate the total delay, this is found experimentally in most cases. This is a critical point as the developer must guarantee that the data rates in all network segments are always kept below the maximum values. Real experiments are presented in the last section showing realistic high frequency control loops and the resulting performance.

As far as software design at the application layer is concerned, a custom protocol is used on top of UDP, for a node and a ROS computer to exchange sensor measurements and control outputs. This includes a byte for defining the message ID, and a number of bytes of data depending on the message ID, see Table 6.1. The protocol is easily adjustable to other applications, as will be shown in the example cases.

6.2.2.13 Software design for the ROS computer

In system design, there is always a discussion concerning the software tool chains that should be used. Also with most platforms, the software created for specific hardware components cannot be transferred easily and reliably to a version consisting of different parts. On the other hand, ROS is a platform including a large database of drivers for devices and sensors, allowing for easy cross communication between processes; this way the need for custom software to handle communication is eliminated. ROS runs on Linux, thus allowing for running the same software on most computers in a technology lab. Also, software for common applications, such as motor control can be used off-the shelf with minimum modifications.

Communication between processes is one of the first challenges a developer faces when designing a robot. ROS provides a messaging system that manages all the communication details, eliminating this way the need for setting up communication protocols, defining data exchange rates etc. Specifically, ROS's basic communication system is an anonymous, asynchronous publish/subscribe mechanism, using nodes (the ROS form of executable files, written in C++, Python or other). A node can publish messages on a bus called topic, to which other nodes can subscribe and receive them. This organization leads to less complex and more readable code. ROS also includes other communication structures like services and actions that can be used depending on the case.

Those said, a ROS running computer is selected here to play the role of the master of the NCS shown in Figure 6.6. In this generic case, the application layer software in the ROS computer side needs to be designed such as to control n motors by exchanging messages with n MCUs, as defined by the protocol described in Table 6.1. To this end, two ROS nodes are required for each motor's low-level control, and a third node to play the coordinating role of a higher level controller.

The first node handles the UDP communication with an MCU; it receives encoder measurements and transmits control outputs, i.e. PWM signal values, back to the MCU. It also communicates with the nodes that implement the controller, the user interface etc. In this interface node, an asynchronous server is set up to receive and send packets, and the IP and the send/receive ports are also defined. The incoming encoder data are read by the node using a *sigaction* function. Every time it receives a message, the signal handler is triggered, the execution of the main function is interrupted, the data are read, and then the execution of the main function continues from where it stopped. Next, the encoder value is published to a topic usually named */state*, as a way of sharing the data with other nodes, e.g. the controller node. Depending on the case, publishing can be done in two ways. If it is done inside the signal handler, the data are published at the same rate they are received, and thus the MCU transmission rate determines also the publishing rate of the node. Alternatively, publishing can be done inside the main function. In this way, and since ROS allows the user to set the loop rate of the main function, the publishing rate is determined by this predefined loop rate and is different from the MCU's transmission rate. Actually, the achieved rate of the node may deviate from the predefined one, since it also depends on the transmission rate of the MCU, the computer resources, the number of running processes and other factors.

Except for sending data, the interface node also needs to receive data from other nodes; by subscribing to a topic named */control_effort*, it reads the calculated by the controller node PWM duty cycle and sends it back to the MCU via UDP. This second controller node operates at the rate it receives messages, i.e. at the rate the */state* topic is published. The publishing rate of the */state* topic is in fact the rate of the control loop. Finally, a third node is typically used as a higher level controller, e.g. to coordinate n actuators to perform a task.

Great attention must be paid to the data rates defined in software design. As can be seen in Figure 6.6, given that n motors are controlled at the same time by the ROS computer, the segment connecting this computer with the switch is where the maximum traffic will appear. Consider a message transmission rate f_{transm} defined for each MCU and the maximum achievable data rate for a segment f_{max} as defined in the previous paragraphs, then for n MCUs the inequality $f_{\text{transm}} < f_{\text{max}}/n$ must always hold to ensure good system performance.

A last critical point concerning the software design of the ROS computer is the processing delay introduced in the total control loop delay, e.g. it is big if graphics are running or data are printed. Also, the non-real time character of the Linux-ROS system gives a non-deterministic character to the system, and thus attention must be paid to the number and the kind of processes running each time. Graphical user interfaces or similar tasks should better run on a different ROS computer connected to the network; ROS allows for easily distributing the application nodes to run in multiple computers.

6.3 Application Experiments

The methods discussed theoretically in the previous sections are applied here to real applications. Two examples concerning legged robotics are presented. The first describes the control system of an instrumented treadmill, on which a single actuated hopping robot – presented as the second application example – can be tested.

6.3.1 Treadmill Control

6.3.1.1 System description
The first example concerns the velocity control of a treadmill (Figure 6.8) placed in the Control Systems Lab (CSL) of the National Technical University of Athens. It is 6 m long and driven by two 3-phase induction

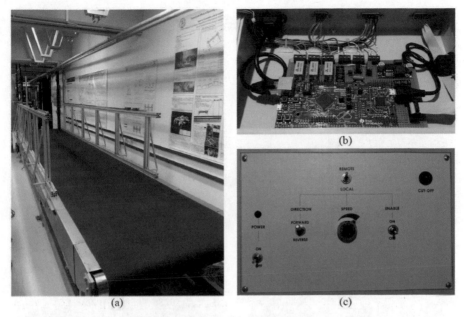

Figure 6.8 (a) The treadmill placed in the Control Systems Lab (CSL) of the NTUA, (b), (c) The treadmill's control system.

motors. The first motor (model MS 100L 2-4, XIUSHI) drives the belt's main pulley achieving a maximum running velocity of 12.6 m/s. The second motor (model FC80-4, Electro Adda) actuates on an endless screw and a rack-pinion system that regulates the treadmill's inclination. Both are driven by inverters; an EMERSON M200-022 model for the belt's motor and a SIEMENS SINAMICS G110 for the inclination motor. These control the motors' velocity according to the formula $n = 120f/P$, where f is the AC current frequency and P the number of poles.

Motor control can be achieved using the control inputs (terminals in Figure 6.9) or the control panel provided by the inverters. While in terminal control mode, at minimum three connections are required (Figure 6.10): system activation through the enable terminal 11, selection of rotation direction (terminals 12 or 13 for forward or reverse respectively) and reference input voltage (terminals 1–2). To drive the motor, both direction and enable terminals have to be connected. With these connections established the motor runs in a minimum frequency predefined in its memory (7–8 Hz). To make the system controllable the terminals need to be connected/disconnected electronically. As the manufacturer specifies, they require a 24-V input, which

Terminal Layout

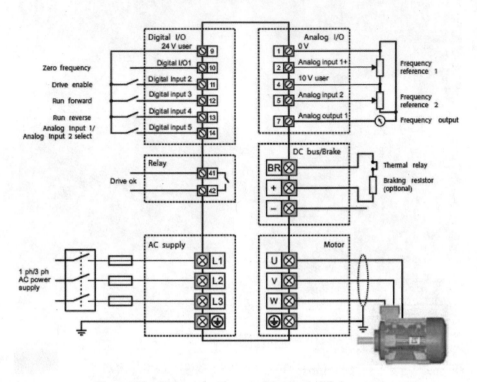

Figure 6.9   The terminal layout of the treadmill's inverter.

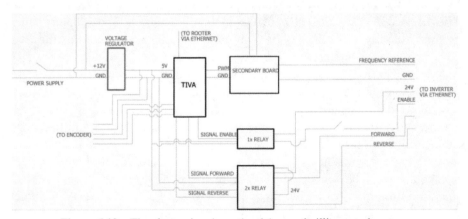

Figure 6.10   The electronic schematic of the treadmill's control system.

can be provided by terminal 9; one needs to short-circuit terminals 9–11 and 9–12 or 9–13 according to the desired direction of motion. To define the frequency, a 0- to10-V input is required on terminal 1 and GND on terminal 2. Three relay modules are used for this, which require a 5-V power supply and a logical signal. To measure the speed of the belt an HEDL-5540 incremental encoder is installed.

To satisfy all requirements, the basic components used are a TM4C1294 Connected LaunchPad, a TP-LINK AC1750 router, and a regular PC running Ubuntu 16.04 and ROS Kinetic. With this setup the treadmill can be controlled also by any other PC on the network. The system built is shown in Figure 6.8.

6.3.1.2 Software design: MCU side
On the MCU side, one needs to enable the QEI and PWM modules and to set up the UDP communication. Moreover, the IP addresses of the MCU and the ROS computer, the device subnet, the device gateway and the send and receive ports must be defined. The software is available on [33].

6.3.1.3 Software design: ROS side
The software designed for the ROS computer consists of three nodes. The first is named *ros_speed_enable* and it is the interface between ROS and the MCU. It sets up the UDP communication, receives velocity measurements from the

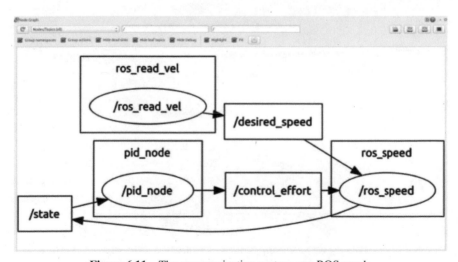

Figure 6.11 The communication system as a ROS graph.

encoder, and publishes them on the topic named */state*. Also, it subscribes on the topic named */control_effort* to receive the calculated PWM commands, which are sent back to the MCU. Note that the send port of the MCU is the receive port for ROS and vice versa. The last topic is published by the control node *pid_node*, which receives the measured velocity from the */state* topic and the desired velocity from the */setpoint* topic as published by the third node *ros_read_velocity*. The latter is the user interface reading the desired speed from the keyboard. The software structure is shown in Figure 6.11 and the software is available on [33].

6.3.1.4 Hardware experiment

An experiment was conducted, where the treadmill was commanded to follow a velocity profile rising linearly with time to 10 m/s, then reducing to 6 m/s, rising again to 10 m/s and finally slowing down back to 0 m/s (Figure 6.12. The sampling frequency was set at 10 kHz, the PID control node frequency was set at 2 kHz, and the velocity was estimated with a frequency of 15 Hz, which however worked well. Also, adjustments of the gains K_p and K_d had to be made.

As deduced from the results shown in Figure 6.12, this is a particularly slow system mainly because of the inverter, which applies a fixed acceleration rate for security reasons; it takes about 22 s to reach 10 m/s. It also seems that the controller needs modifications to reduce the high-frequency oscillations that appear.

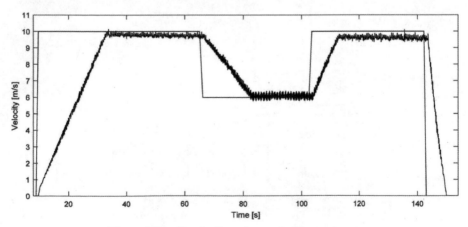

Figure 6.12   Results from the treadmill experiment.

6.3.2 Single Actuated Hopping Robot (SAHR)

This example refers to the Single Actuated Hopping Robot (SAHR), designed and built in CSL [34]. Practically, it is a realization of the standard Spring Loaded Inverted Pendulum (SLIP) model. Here, the hardware and software are redesigned based on the principles presented in the previous sections.

6.3.2.1 Robot description

The system is designed to move on the sagittal plane and has an actuated revolute hip joint, and a passively compliant prismatic knee joint using a spring of stiffness 5900 N/m. Most parts are made of aluminum, while the leg shaft receiving most of the impact loads is made of steel. A Maxon motor (RE35, 90 W) is used with maximum continuous current 3.36 A, nominal voltage 24 V and maximum continuous torque 0.0933 Nm. A planetary gearhead with gear ratio 26:1 is attached together with a belt drive of reduction ratio 2:1 to move the hip axis. The motor drive is an AZBDC12A8 by Advanced Motion Controls (AMC), supplying up to 6 A continuous and 12 A intermittent current. To use it, one has to send an enabling signal to the corresponding pin, a High/Low signal to the direction pin, and a PWM signal corresponding to the desired current (by default, 100% duty cycle corresponds to 12 A).

Regarding the sensory system, two incremental encoders measure the spring compression and the hip angle, providing three pulses, A, B, and Index. Measurements from the spring encoder can be used to calculate the body CoM position in stance phase, detect transitions from stance to flight phase and vice versa, or even estimate the ground force. However, for a better ground force measurement, a 3-Axis force sensor by BOTA Systems is also installed.

6.3.2.2 Software design: MCU side

Two TM4C1294 boards are used to read the sensors since each board has only one QEI module. The board that reads the hip encoder also sends commands to the motor drive. Their programming is identical to that for the treadmill control system, with small modifications concerning the IP addresses and the encoder related parameters. The software is available on [33].

6.3.2.3 Software design: ROS side

In this case, where two encoders need to be read, the advantages of ROS become apparent. Two identical MCU interface nodes are used for the

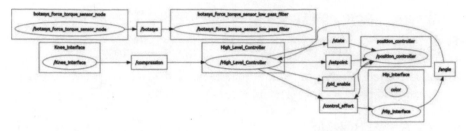

Figure 6.13 The ROS graph for the control system of SAHR.

encoder measurements to become available for every node that needs them. To control the hip angle a PID controller node from the *ros_pid* package [35] is used. The force sensor connected to the computer is ROS-ready and only the driver needs to be built. Extra nodes can be also added depending on the application. For a position control experiment a user interface is required to send commands, and for a force sensing experiment the force sensor driver must run. In the case of a hopping experiment, a high level controller is also required to read the knee encoder or the force sensor data and decide if the leg is in stance or flight phase (Figure 6.13). In flight phase it can position the leg for landing, while in stance it can push the body forward with a predefined torque. However, more complex control algorithms can be used by taking into account the ground stiffness and the energy losses as proposed in Vasilopoulos et al. [36]. The software is available on [33].

6.3.2.4 Simulation experiment

The purpose of the experiment is to determine how close a Gazebo simulation is to reality, but also to test the developed ROS system. A big advantage of the Gazebo and ROS setup is the direct simulation with the software used on the actual hardware. In this case two custom plugins are used; one to publish the joint states and one to subscribe to a topic to receive hip commands (just like the MCUs in the real robot).

The robot was left to fall from a height of 0.1 m, and the leg was controlled to remain vertical. The ground was considered infinitely stiff, and this might be one of the reasons that a small deviation between simulation and experiment was observed. The ODE solver was used with a 0.001 s maximum step size. The simulation results are shown in Figure 6.14 and also in the video presented in SAHR [37].

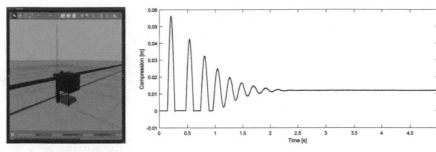

Figure 6.14   Gazebo simulation for the monopod hopper SAHR.

6.3.2.5 Hardware experiment

To constraint the robot to move on the sagittal plane, it is mounted on a supporting mechanism on the treadmill. Like in the simulation, the leg is controlled to stay in the vertical position with a PD controller, as shown in Figure 6.15. Leaving the rest of the ROS system the same, the MCU nodes are added. The transmission frequency of each MCU is set to 15 kHz and the PID frequency to 1 kHz.

Of great importance is the good matching between simulation and experiment for many aspects in robotics research. As deduced from Figure 6.14 and Figure 6.16, simulation and reality were close in this experiment, showing the same number of bounces, the same settling time (about 2.3 s) and the same steady-state compression (0.012 m). A slight difference is observed in the maximum compression appearing higher in simulation, which could be

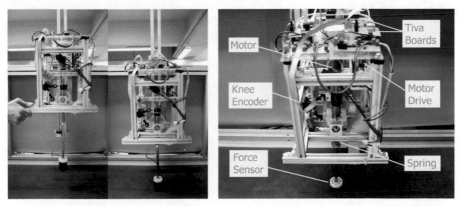

Figure 6.15   The Single Actuated Hopping Robot (SAHR) in a hopping experiment.

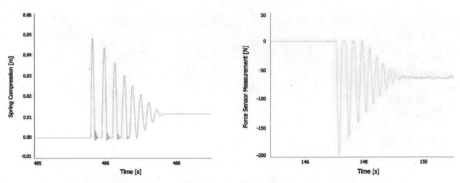

Figure 6.16   Results from the SAHR hopping experiment.

explained by unmodeled frictional losses. A video of the experiment can be found in SAHR [37].

6.3.2.6 Simulations on interactions with terrains

Another interesting example of simulations with SAHR, is the case of planetary environments with different gravitational accelerations and different terrain types. In the following example [38], a controller presented in Vasilopoulos et al. [36], tries to move the SAHR in three different environments. It is interesting to see how the motion profiles are affected by the abovementioned parameters. The robot behavior was simulated for the

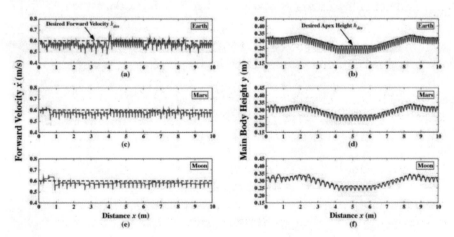

Figure 6.17   Controller performance on shallow crater: (a, c, e) Forward Velocity on Earth, Mars and Moon respectively, (b, d, f) Main Body Height on Earth, Mars and Moon, respectively.

Earth ($g = 9.81$ m/s^2), the Mars ($g = 3.711$ m/s^2) and the Moon ($g = 1.622$ m/s^2). While the controller adapted quickly to each terrain and followed the desired objectives of forward velocity and body height in each case, as it is shown in Figure 6.17, it can be observed that the response converged more slowly to the desired commands as the acceleration of gravity decreased.

6.4 Conclusions

In this chapter, advanced but easy to implement and low-cost methods were proposed for modeling, control, design and development of robotic systems. Widely used software tools such as Matlab, Gazebo, and ROS were combined to form a framework for fast simulation and prototyping of robots seen as Networked Control Systems (NCSs). The focus was on giving guidelines on how to directly test a conceptual idea in design and control using low-cost solutions in software and hardware. Two application examples concerning legged robotics were finally presented as a proof of concept, and all the software was built such as to be easily reused by the reader in similar or modified applications.

Acknowledgment

Funding for this research by the "IKY Fellowships of Excellence for Post-graduate Studies in Greece – Siemens Programme" in the framework of the Hellenic Republic – Siemens Settlement Agreement is acknowledged. The authors wish to thank Stylianos Vagenas for assistance with the treadmill control, and Pavlos Stavrou for assistance with the development of software.

References

[1] Marwedel, P. (2011). *Embedded System Design, Embedded Systems Foundations of Cyber-Physical Systems*. Netherland: Springer.
[2] Rawat, D. B., Rodriques, J., and Stojmenovic, I. (2015). *Cyber Physical Systems: From Theory to Practice*. Boca Raton: CRC Press, Taylor & Francis Group.
[3] Jeschke, S., Brecher, C., Song, H., and Rawat, D. B. (2017). *Industrial Internet of Things, Cybermanufacturing Systems*. Berlin: Springer International Publishing.

[4] Gilchrist, A. (2016). *Industry 4.0: The Industrial Internet of Things*. New York, NY: Apress.

[5] Song, H., Rawat, D., Jeschke, S., and Brecher, C. (2017). *Cyber-Physical Systems: Foundations, Principles and Applications*. Academic Press.

[6] ROS. http://www.ros.org/

[7] Raspberry Pi. https://www.raspberrypi.org/

[8] Ivaldi, S., Padois, V., and Nori, F. (2014). Tools for dynamics simulation of robots: a survey based on user feedback. arXiv preprint. arXiv:1402.7050

[9] https://github.com/kostasmach/Laelaps_sim

[10] Machairas, K., and Papadopoulos, E. (2016). "An Active Compliance Controller for Quadruped Trotting," in *24th Mediterranean Conference on Control and Automation (MED '16)*, Athens, Greece.

[11] Stronge, W. J. (2000). *Impact Mechanics*. Cambridge: Cambridge University Press.

[12] Hunt, K. H. and Crossley, F. R. E. (1975). Coefficient of Restitution Interpreted as Damping in Vibroimpact. *J. Appl. Mech.* 440–445.

[13] Majeed, M. A., Yigit, A. S., and Christoforou, A. P. (2012). "Elasto-plastic contact/impact of rigidly supported composites. *Compos. Part B Eng.*, vol. 43, 1244–1251.

[14] Paraskevas, I. (2015). "Capturing of Orbital Space Systems by Robots." Ph.D. Dissertation, National Technical University of Athens, Greece.

[15] Gupta, R. A., and Chow, M.-Y. (2010). Networked control system: Overview and research trends. *IEEE Trans. Ind. Electr.* 57, 2527–2535.

[16] Valenzano, A., and Cena, G. (2014). "Protocols and Services in Controller Area Networks," in *Industrial Communication Technology Handbook,* 2nd Edn. (Boca Raton, FL:), 52.1–52.49.

[17] Cena, G., Scanzio, S., Valenzano, A., and Zunino, C. (2014). "Ethernet for Control Automation Technology," in *Industrial Communication Technology Handbook, Second Edition* (Boca Raton, FL: CRC Press), 18.1–18.27.

[18] Zurawski, R. (2014). *Industrial Communication Technology Handbook*, 2nd Edn. Boca Raton, FL: CRC Press, 2014.

[19] Sauter, T., Soucek, S., Kastner, W., and Dietrich, D. (2011). The evolution of factory and building automation. *IEEE Ind. Electr. Mag.* 5, 35–48.

[20] Zurawski, R. (2014). "Fieldbus System Fundamentals," in *Industrial Communication Technology Handbook, 2nd Edn*, Boca Raton, FL: CRC Press.

[21] Lounsbury, R. (2008). *Industrial Ethernet on the plant floor: A planning and installation Guide*, ISA.

[22] Kurose, J. F., and Keith, R. W. (2013). *Computer networking: a top-down approach* (6th Edition). Upper Saddle River, NJ: Pearson.

[23] Zurawski, R. (2014). "Networked Control Systems for Manufacturing," in *Industrial Communication Technology Handbook*. Boca Raton, FL: CRC Press.

[24] Axelson, J. (2003). *Embedded Ethernet and Internet Complete*, Lakeview Research LLC.

[25] Enner, F. (2016). *Analyzing the viability of Ethernet and UDP for robot control*. https://ennerf.github.io/2016/11/23/Analyzing-the-viability-of-Ethernet-and-UDP-for-robot-control.html

[26] Prytz, G., and Johannessen, S. (2005). "Real-time performance measurements using UDP on Windows and Linux," in *10th IEEE Conference on Emerging Technologies and Factory Automation, ETFA* (New York, NY: IEEE).

[27] EtherCAT. https://www.ethercat.org/

[28] Zimmerman, J. and Spurgeon, C. E. (2014). *Ethernet: The Definitive Guide*, 2nd Edn. Newton, MA: O'Reilly Media, Inc.

[29] Loeser, J., and Haertig, H. (2004). "Low-latency hard real-time communication over switched Ethernet," in *16th Euromicro Conference on Real-Time Systems, ECRTS* (New York, NY: IEEE).

[30] Parrott, J. T. Moyne, J. R., and Tilbury, D. M. (2006). "Experimental Determination of Network Quality of Service in Ethernet: UDP, OPC, and VPN," in *American Control Conference*, Minneapolis, Minnesota, USA.

[31] Gottlieb, I. M. (1993). *Practical RF power design techniques*. New York, NY: TAB Books.

[32] EK-TM4C1294. http://www.ti.com/tool/ek-tm4c1294xl

[33] https://github.com/ntua-cslep/Legged

[34] Cherouvim, N., and Papadopoulos, E. (2008). "The SAHR robot: Controlling Hopping Speed and Height Using a Single Actuator," in *Applied Bionics and Biomechanic,* vol. 5, 149–156.

[35] http://wiki.ros.org/pid

[36] Vasilopoulos, V., Paraskevas, I., and Papadopoulos, E. (2015). "Control and Energy Considerations for a Hopping Monopod on Rough Compliant Terrains," in *IEEE International Conference on Robotics and Automation (ICRA '15)*, Seattle, WA, USA.

[37] SAHR. (2016). *Single Actuated Hopping Robot experiment.* https://www.youtube.com/watch?v=Dcw0f3NULy4&feature=youtu.be

[38] Vasilopoulos, V. Paraskevas, I., and Papadopoulos, E. (2015). "Monopod Hopping on Rough Planetary Environments," in *13th Symposium on Advanced Space Technologies in Robotics and Automation, (ASTRA '15)*, ESA, ESTEC, Noordwijk, The Netherlands.

7

Modeling Control Mechanisms in MATLAB

Kostas Siozios

Department of Physics, Aristotle University of Thessaloniki, Greece

Abstract

This chapter presents a number of fundamental functions required for the decision-making, i.e., control, of cyber-physical systems. For this purpose, the MATLAB software tool is employed.

7.1 Introduction

MATLAB is an integrated environment in which engineers can model, simulate, and design control systems. By combining mathematics functionality and comprehensive engineering with powerful visualization and animation features, all within a high-level interactive programming language, it is feasible to derive application-specific customizations on the decision-making mechanisms, which in turn results to maximizing the overall system's performance metrics. Additionally, the availability of MATLAB toolboxes extends further the MATLAB environment to incorporate a wide range of classical and modern techniques for the design of control systems, providing cutting edge control algorithms developed by internationally recognized experts. Current version of MATLAB suite contains more than 600 mathematical, statistical and engineering functions, providing the power of numerical calculation you need to analyze data, develop algorithms and optimize the performance of a system. In addition, MATLAB is a high-level programming language that allows you to develop algorithms in a fraction of the time spent in C, C++, or FORTRAN.

Within this chapter, we provide a number of case studies, where fundamental techniques of decision-making mechanisms are employed. These techniques usually are the core technology not only for designing control mechanisms of complex cyber-physical systems, but they are also useful for comparing performance of alternative existing control strategies.

As discussed in previous chapters, various layers of abstraction can be employed for this purpose. More precisely, the diagram in Figure 7.1 shows how an engineering problem leads to the development of models and the analysis of experimental data, which in turn lead to the design and simulation of control systems. The subsequent analysis of these systems leads to further modifications of the design, this development loop resulting in rapid prototyping and implementation of effective systems.

The interactive *Control System Toolbox* tools facilitate the design and adjustment of control systems. For example, you might drag poles and zeros and see immediately how the system reacts. In addition, MATLAB provides powerful interactive 2-D and 3-D graphics features showing data, equations, and results. It is possible to use a wide range of visualization aids in MATLAB or you can take advantage of the specific control functions which are provided by the MATLAB toolboxes.

The MATLAB toolboxes include applications written with MATLAB language-specific functionality. The MATLAB control-related toolboxes encompass virtually all of the fundamental techniques of control design, from LQG and root-locus to H and logical diffuse methods. For example, it might

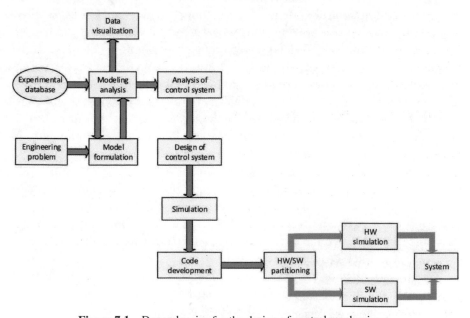

Figure 7.1   Dependencies for the design of control mechanisms.

add a fuzzy logic control system design using the built-in algorithms of the *Fuzzy Logic Toolbox*.

The most important MATLAB toolboxes for control systems can be classified into three families: modeling (*System Identification Toolbox*), classical design and analysis products (*Control System Toolbox* and *Fuzzy Logic Toolbox*), design and advanced analysis products (*Robust Control Toolbox, Mu-Analysis Toolbox, LMI Control Toolbox* and *Model Predictive Toolbox*), and optimization products (*Optimization Toolbox*). The rest of this chapter pre-assumes that readers are familiar with basic knowledge of MATLAB programing.

7.2 MATLAB Control System Toolbox

The control system toolbox provided by MATLAB suite is useful in order to enable easily tuning of PID controllers. PID control is not a trivial task. While simple in theory, design, customization, and implantation of PID controllers can be difficult and time-consuming in the majority of usecases.

More precisely, the PID control involves several tasks that include among others:

- The selection of an appropriate PID algorithm (P, PI, or PID)
- The tuning of controller gains
- The simulation of designed controller against a plant model
- The implementation and customization of the controller on a target processing unit.

Since the previously mentioned tasks are complex enough, scientific software tools, such as MATLAB suite, are absolutely necessary in order to support different tasks of this procedure. The advantages of using such an approach includes:

- Configure your Simulink PID Controller block for PID algorithm (P, PI, or PID), controller form (parallel or standard), anti-windup protection (on or off), and controller output saturation (on or off).
- Support both the automatic controller tuning, as well as the fine-tune process interactively.
- Since a typical system contains multiple control units, all of them can be tuned in batch mode.
- Run closed-loop system simulation by connecting your PID controller block to the plant model
- Automatically generate C code for targeting a microcontroller.

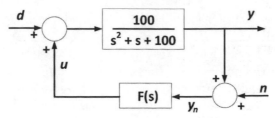

Figure 7.2   Design of a Linear–Quadratic–Gaussian (LQG) controller.

Additionally, the majority of these software tools include also a number of toolboxes, each of which solves dedicated problems of system design. For instance, the MATLAB suite has the *Control System Toolbox*, which supports the representation of four linear models:

- state–space models (SS);
- transfer functions (TF);
- zero-pole-gain models (ZPK);
- frequency data models (FRD).

Linear time-interval (LTI) objects are provided for each model type. In addition to model data, LTI objects can store the sample time of discrete-time systems, delays, names of inputs and outputs, notes on the model and many other details. Using LTI objects, an engineer can manipulate models as unique entities and combine them using matrix-type operations.

An illustrative example of the design of a simple Linear–Quadratic–Gaussian (LQG) controller is shown in Figure 7.2. The code extract at the bottom shows how the controller is designed and how the closed-loop system has been created. The plot of the frequency response shows a comparison between the open-loop system (red) and closed loop system (blue).

Next, we provide the MATLAB source code for this case:

```
%% Matlab script
%% state-space plant model
>> G = ss(tf(100, [ 1  1  100 ]));

% design feedback gain matrix
>> Klqr = lqry( G, 10, 1 );

% Kalman estimator design
>> Kest = kalman(G(: , [1 1]),  1,  0.01);

% combine regulator and estimator
>> F = lqgreg( Kest, Klqr );
```

```
% form closed-loop system
>> clsys = feedback (G, F, +1);

% generateand plot impulse response
>> impulse(G, 'r', clsys, 'b');
```

The output of this program follows. Red color line
corresponds to the frequency response between the
open-loop system, whereas blue color line refers to
the closed loop system.

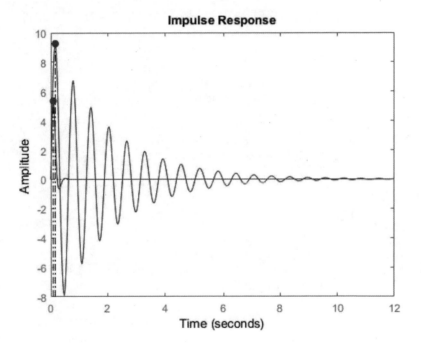

7.3 Overview of Commands for the Control System Toolbox

The *Control System Toolbox* contains commands which analyze and compute
model features such as I/O dimensions, poles, zeros and DC gain. These
commands apply both to continuous-time and discrete-time models. This
section summarizes the main commands of the Control System Toolbox,
which can be classified according to their purpose as follows:

General

Ctrlpref	Opens a GUI which allows you to change the *Control System Toolbox* preferences

Creation of linear models

tf	Creates a transfer function model
zpk	Creates a zero-pole-gain model
ss	Creates a state–space model
dss	Creates a descriptor state–space model
frd	Creates a frequency–response data model
set	Locates and modifies properties of LTI models

Data extraction

tfdata	Accesses transfer function data (in particular extracts the numerator and denominator of the transfer function)
zpkdata	Accesses zero-pole-gain data
ssdata	Accesses state–space model data
get	Accesses properties of LTI models

Conversions

s	Converts to a state–space model
zpk	Converts to a zero-pole-gain model
tf	Converts to a transfer function model
frd	Converts to a frequency–response data model
c2d	Converts a model from continuous to discrete time
d2c	Converts a model from discrete to continuous time
d2d	Resamples a discrete time model

System interconnection

append	Groups models by appending their inputs and outputs
parallel	Parallel connection of two models
series	Series connection of two models
feedback	Connection feedback of two systems
lft	Generalized feedback interconnection of two models
connect	Block diagram interconnection of dynamic systems

Dynamic models

iopzmap	Plots a pole-zero map for input/output pairs of a model
bandwidth	Returns the frequency-response bandwidth of the system
pole	Computes the poles of a dynamic system
zero	Returns the zeros and gain of a SISO (single-input single-output) dynamic system
pzmap	Returns a pole-zero plot of a dynamic system
damp	Returns the natural frequency and damping ratio of the poles of a system
dcgain	Returns the low frequency (DC) gain of an LTI system
norm	Returns the norm of a linear model
covar	Returns the covariance of a system driven by white noise

Time-domain analysis

ltiview	An LTI viewer for LTI system response analysis
step	Produces a step response plot of a dynamic system
impulse	Produces an impulse response plot of a dynamic system
initial	Produces an initial condition response plot of a state–space model
lsim	Simulates the time response of a dynamic system to arbitrary inputs

Frequency-domain analysis

ltiview	An LTI viewer for LTI system response analysis
bode	Produces a Bode plot of frequency response, magnitude and phase of frequency response
sigma	Produces a singular value plot of a dynamic system
nyquist	Produces a Nyquist plot of frequency response
nichols	Produces a Nichols chart of frequency response
margin	Returns gain margin, phase margin, and crossover frequencies
allmargin	Returns gain margin, phase margin, delay margin and crossover frequencies
freqresp	Returns frequency response over a grid

Classic design

sisotool	Interactively design and tune SISO feedback loops (technical *root locus* and *loop shaping*)
rlocus	Root locus plot of a dynamic system

Pole placement

place	MIMO pole placement design
estim	Forms a state estimator given estimator gain
reg	Forms a regulator given state-feedback and estimator gains

LQR/LQG design

lqr	Linear quadratic regulator (LQR) design
dlqr	Linear–quadratic (LQ) state-feedback regulator for a discrete-time state-space system
lqry	Linear–quadratic (LQ) state-feedback regulator with output weighting
lqrd	Discrete linear–quadratic (LQ) regulator for a continuous plant
Kalman	Kalman estimator
kalmd	Discrete Kalman estimator for a continuous plant

State–space models

rss	Generates a random continuous test model
drss	Generates a random discrete test model
ss2ss	State coordinate transformation for state–space models
ctrb	Controllability matrix
obsv	Observability matrix
gram	Control and observability gramians
minreal	Minimal realization or pole-zero cancelation
ssbal	Balance state–space models using a diagonal similarlity transformation
balreal	Gramian-based input/output balancing of state–space realizations
modred	Model order reduction

Models with time delays

totaldelay	Total combined input/output delay for an LTI model
delay2z	Replaces delays of discrete-time TF, SS, or ZPK models by poles at $z=0$, or replaces delays of FRD models [Note: in more recent versions of MATLAB, *delay2z* has been replaced with *absorbDelay*.]
pade	Pade approximation of a model with time delays

Matrix equation solvers

lyap	Solves continuous-time Lyapunov equations
dlyap	Solves discrete-time Lyapunov equations
care	Solves continuous-time algebraic Riccati equations
dare	Solves discrete-time algebraic Riccati equations

Commands for Interconnecting Models

Combines models in a diagonal configuration block. Groups the models together by appending their inputs and outputs (see Figure 7.3).	*sys = append(sys1, sys2, ..., sysN)*
Appends the state vector to the output vector.	*asys = augstate (sys)*
Connects the subsystems in a block according to a chosen interconnection scheme (given by the connection matrix Q).	*sysc = connect(sys, Q, inputs, outputs)*
Returns a model sys for the negative feedback interconnection of models sys1 and sys2 (see Figure 7.4). May include sign and closed loop (see Figure 7.5).	*sys = feedback(sys1, sys2)* *sys = feedback(sys1, sys2, sign)* *sys =feedback(sys1, sys2, feedin, feedout, sign)*
Forms the linear fractional transformation (LFT) of two models (see Figure 7.6).	*sys = lft(sys1, sys2)*
Generates continuous second-order systems (*wn* is the natural frequency and *z* is the damping factor).	*sys = lft(sys1, sys2, nu, ny)* *[A, B, C, D] = ord2(wn, z)*
Connects two systems in parallel (see Figure 7.7).	*[num, den] = ord2(wn, z)* *sys = parallel(sys1, sys2)*
Connects two systems in series (see Figure 7.8).	*sys =parallel(sys1, sys2, inp1, inp2, out1, out2)* *sys = series(sys1, sys2)*
Produces an array of dynamic system models by stacking the models sys1, sys2, ... along the array dimension arraydim.	*sys = series(sys1, sys2, out1, inp2)* *sys = stack(arraydim, sys1, sys2, ...)*

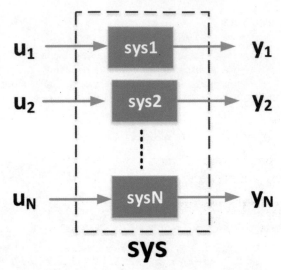

Figure 7.3   Models in a diagonal configuration block.

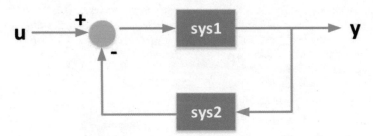

Figure 7.4   Negative feedback interconnection.

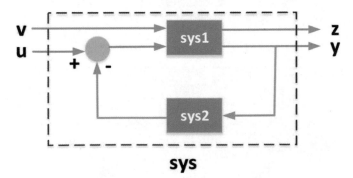

Figure 7.5   Negative feedback interconnection (alternative topology).

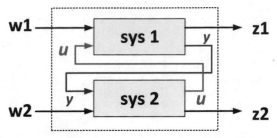

Figure 7.6 Linear fractional transformation (LFT) of two models.

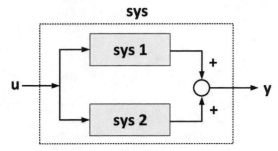

Figure 7.7 Two systems in parallel.

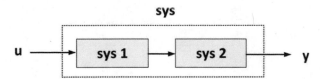

Figure 7.8 Two systems in series.

7.4 Examples of Designing Control Systems at MATLAB

Next, we provide a number of examples, each of which deals with basic concepts of designing control systems.

7.4.1 Example 7.1

Determine the pole locations for the system shown below using MATLAB.

$$\frac{C(s)}{R(s)} = \frac{2s^3 - 4s^2 + 3s + 2}{s^5 - 2s^4 + 3s^3 + s^2 - 4s + 1}$$

Solution:

```
%% Matlab script
>> den= [ 1   -2   3   1   -4   1 ];
>> B = rots(den);

B =
     0.8576 + 1.6661i
     0.8576 - 1.6661i
    -1.0000 + 0.0000i
     1.0000 + 0.0000i
     0.2848 + 0.0000i
```

7.4.2 Example 7.2

Determine the pole locations for the unity feedback system shown below using MATLAB.

$$G\left(s\right) = \frac{120}{\left(s+3\right)\left(s+4\right)\left(s+1\right)\left(s+5\right)}$$

Solution:

```
%% Matlab script
» numg = 120

numg =

       120

» deng = poly ([ -3 -4 -1 -5 ])

deng =
       1      13      59      107      60

» 'G(s)'

ans =

G(s)
```

```
» G=tf(numg, deng)

G =
                      120
    ------------------------------------------------
      s^4 + 13 s^3 + 59 s^2 + 107 s + 60

Continuous-time transfer function.

» 'Poles'

ans =

Poles

» pole(G)

ans =
        -5.0000
        -4.0000
        -3.0000
        -1.0000

» 'T(s)'

ans =

T(s)

» T=feedback(G, 1)

T =
                      120
    ----------------------------------
      s^4 + 13 s^3 + 59 s^2 + 107 s + 180

Continuous-time transfer function.

» pole(T)

ans =
      -5.8168 + 2.1410i
```

```
-5.8168 - 2.1410i
-0.6832 + 2.0539i
-0.6832 - 2.0539i
```

7.4.3 Example 7.3

A plant to be controlled is described by the following transfer function. Compute the root locus plot using MATLAB.

$$G(s) = \frac{s^2 + 3s + 2}{s^3 + 2s^2 + 3s + 4}$$

Solution:

```
%% Matlab script
» clf;
» num = [ 1 3 2 ];
» den = [ 1 2 3 4 ];
» rlocus ( num, den);

Computer response is shown in upcoming figure.
```

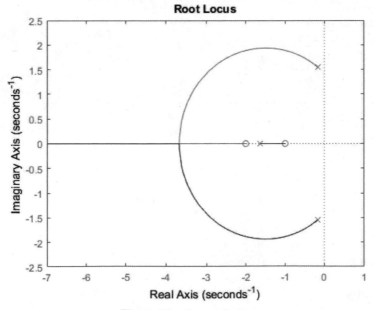

Figure 7.9   Output figure.

7.4.4 Example 7.4

For the unity feedback system shown below, G(s) is given as

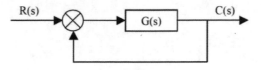

$$G\,(s) = \frac{15(s^2 + 3s + 7)}{(s^2 + 3s + 7)(s+1)(s+3)}$$

Determine the closed-loop step response using MATLAB.
Solution:

```
%% Matlab script
» numg = 15*[ 1 3 7 ];
» deng = conv([ 1 3 7 ], poly([ -1 -3 ]));
» G = tf(numg, deng);
» T = feedback(G, 1);
» step(T)

Next figure plots the response function:
```

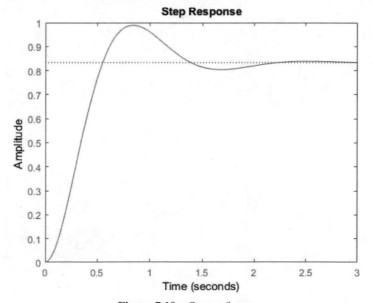

Figure 7.10 Output figure.

7.4.5 Example 7.5

Write a program in MATLAB to obtain the Nyquist plot for the following transfer function for $k = 30$.

$$G(s) = \frac{k(s+3+7\mathrm{i})(s+1)(s+3-7\mathrm{i})}{(s+5)(s+3)(s+1)(s+3-7\mathrm{i})(s+3+7\mathrm{i})}$$

Solution:

```
%% Matlab script
» %Simple Nyquist and Nichols plots
» clf
» z = [ -3-7*i -1 -3+7i ];
» p = [ -5 -3 -1 -3+7i -3-7*i ];
» k = 30;
» [num, den] = zp2tf (z', p', k);
» plot (211), nyquist (num, den)

The output Nyquist diagram follows:
```

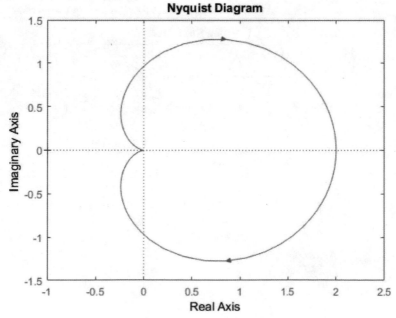

Figure 7.11   Output figure.

7.4.6 Example 7.6

A PID controller is given by

$$G(s) = 29.125 \times \frac{(s^2 + 1.14s + 0.3249)}{s}$$

Draw a Bode diagram of the controller using MATLAB.

Solution:

The equation is written as follows:

$$G(s) = \frac{29.125s^2 + 33.2025s + 9.4627}{s}$$

The following MATALB program produces the Bode diagram

```
%% Matlab script
» %Bode diagram
» num = [ 29.125 33.2025 9.4627 ];
» den= [ 0 1 0];
» bode (num, den)
» title ('Bode diagram of G(s)')
```

The output Bode diagram of the controller follows:

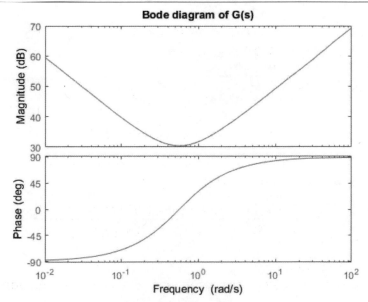

Figure 7.12   Output figure.

7.4.7 Example 7.7

Write a program in MATLAB for the system shown below so that the value of K can be input (K = 35).

$$\frac{C(s)}{R(s)} = \frac{K(s+2)(s+3)}{s(2s^2 + 10s + 3)}$$

(a) Display the closed-loop magnitude and phase frequency response for unity feedback system with an open-loop transfer function, $KG(s)$.

(b) Determine and display the peak magnitude, frequency of the peak magnitude, and bandwidth for the closed-loop frequency response for the input value of K.

Solution:

```
%% Matlab script
» numg = [ 0 0 2 3 ];
» deng = [ 1 2 10 3 ];
» 'G(s)'
» G = tf(numg, deng)
» K = input( 'Type gain, K = ' );
» 'T(s)'
» T = feedback(K*G, 1)
» bode(T)
» title('Closed-loop frequency response')
» [M, P, w] = bode(T);
» [Mp i] = max(M);
» Mp
» MpdB = 20*log10(Mp)
» wp = w(i)
» for i = 1:1:length(M);
» if M(i) <= 0.707;
» fprintf('Bandwidth = %g', w(i))
» break
» end
» end

The output follows:

ans =
```

```
G(s)


G =

            2 s + 3
     -------------------------------
      s^3 + 2 s^2 + 10 s + 3

Continuous-time transfer function.
```

Type gain, K = 35

```
ans =

T(s)


T =

             70 s + 105
     ------------------------------------
      s^3 + 2 s^2 + 80 s + 108

Continuous-time transfer function.

Mp =

     12.4256


MpdB =

     21.8863

wp =
      8.8843

Bandwidth = 13.3645

The output figure follows:
```

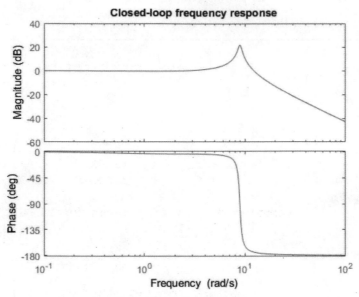

Figure 7.13   Output figure.

7.4.8 Example 7.8

Determine the unit-ramp response of the following system using MATLAB and *lsim* command.

$$\frac{C(s)}{R(s)} = \frac{s+2}{3s^2 + 6s + 2}$$

Solution:

```
%% Matlab script
>> %Unit-ramp response
>> num = [ 0   1   2 ];
>> den = [ 3   6   2 ];
>> t = 0:0.1:8;
>> r = t;
>> y = lsim(num, den, r, t);
>> plot(t, r,   '-', t, y, 'o')
>> grid
>> title('Unit-ramp response')
>> xlabel('t Sec')
```

```
>> ylabel('Unit-ramp input and output')
>> text(1.0,   4.0,   'Unit-ramp input')
>> text(5.0,   2.0,   'Output')
```

The output follows:

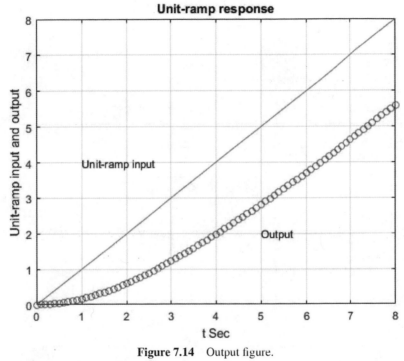

Figure 7.14   Output figure.

7.4.9 Example 7.9

A unity-feedback control system is defined by the following feedforward transfer function:

$$G(s) = \frac{K(s+3)}{s(s^2 + 2s + 9)}$$

(a) determine the location of the closed-loop poles, if the value of gain is equal to 4.
(b) plot the root loci for the system using MATLAB.

Solution:

```
%% Matlab script
%% find the closed-loop poles
>> p = [1 5 9 4];
>> roots(p)

ans =
    -2.1766 + 1.2028i
    -2.1766 - 1.2028i
    -0.6468 + 0.0000i

>> %MATLAB Program to plot the root-loci
>> num = [ 0   0   1   3 ];
>> den = [ 1   2   9   0 ];
>> rlocus(num, den);
>> grid
>> title('Root-locus plot of G(s)')

The root loci for the system follows:
```

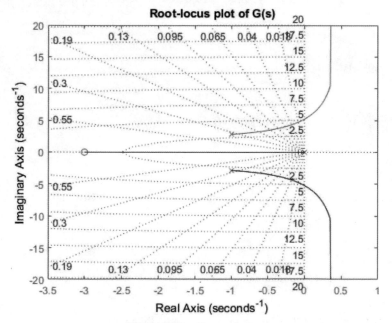

Figure 7.15   Output figure.

7.4.10 Example 7.10

The open-loop transfer function of a unity-feedback control system is given by

$$G(s) = \frac{s+2}{s^3 + 3s + 2}$$

(a) draw a Nyquist plot of G(s) using MATLAB
(b) determine the stability of the system.

Solution:

```
%% Matlab script
>> % Open-loop poles
>> p = [1  0  3  2];
>> roots(p)

ans =

        0.2980 + 1.8073i
        0.2980 - 1.8073i
       -0.5961 + 0.0000i



>> % Nyquist plot
>> num = [ 0    0    1    2  ];
>> den = [ 1    0    3    2  ];
>> nyquist(num, den)
>> grid
>> title('Nyquist plot of G(s)')

The Nyquist plot of G(s) follows:

There are two open-loop poles in the right half s plane
and no encirclement of the critical point, the closed-
loop system is unstable.
```

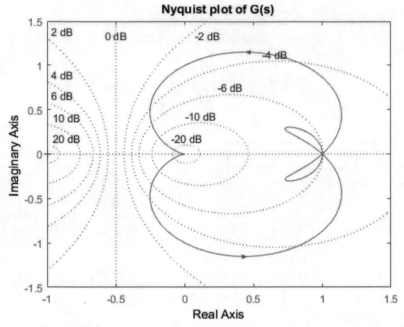

Figure 7.16   Output figure.

7.4.11 Example 7.11

The open-loop transfer function of a unity-feedback control system is given by

$$G\left(s\right) = \frac{K\left(4s + 2\right)}{s^3 + 5s^2 + 3s}$$

Plot the Nyquist diagram of $G(s)$ for $K = 1$, 10, and 100 using MATLAB.
Solution:

```
%% Matlab script
» num = [ 4 2 ];
» den = [ 1 5 3 0 ];
» w = 0.1:0.1:100;
» [re1, im1, w] = nyquist(num, den, w);
» [re2, im2, w] = nyquist(10*num, den, w);
» [re3, im3, w] = nyquist(100*num, den, w);
» plot(re1, im1, re2, im2, re3, im3)
```

```
» v = [ -3 3 -3 3 ];
» axis(v);
» grid
» title('Nyquist diagrams')
» xlabel('Real axis')
» ylabel('Imaginary axis')
» text(-0.2, -2, `K = 1')
» text(-1.5, -2.0, `K = 10')
» text(-2, -1.5, `K = 100')
```

The Nyquist diagrams of $G(s)$ for K = 1, 10, and 100 follows:

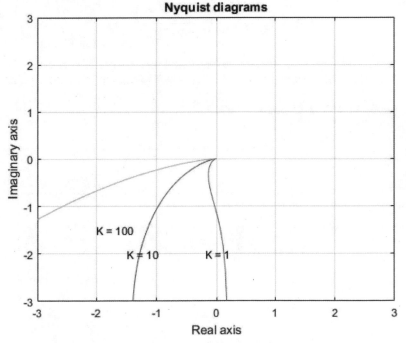

Figure 7.17   Output figure.

7.4.12 Example 7.12

For the closed-loop control system shown in upcoming figure, obtain the range of gain K for stability and plot a root-locus diagram for the system.

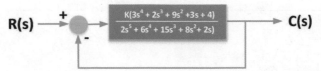

Figure 7.18   Control system for Example 12.

Solution:

The range of gain K for stability is obtained by
first plotting the root loci and then finding critical
points (for stability) on the root loci. The open-loop
transfer function G(s) is

$$G\left(s\right) = \frac{K(3s^4 + 2s^3 + 9s^2 + 3s + 4)}{2s^5 + 6s^4 + 15s^3 + 8s^2 + 2s}$$

A MATLAB program to generate a plot of the root
loci for the system is given below.

```
%% Matlab script
» num = [ 0 3 2 9 3 4 ];
» den = [ 2 6 15 8 2 0 ];
» rlocus(num, den);
» v = [ -1.5 0 -3 3];
» axis(v);
» axis('square');
» grid;
» title('Root-Locus Plot');
```

The resulting root-locus plot is shown in next figure:

7.4.13 Example 7.13

A control system is defined by the following state space equations:

$$\begin{bmatrix} \dot{x}_1 \\ \dot{x}_2 \end{bmatrix} = \begin{bmatrix} -4 & -1 \\ 2 & -3 \end{bmatrix} \begin{bmatrix} x_1 \\ x_2 \end{bmatrix} + \begin{bmatrix} 1 \\ 3 \end{bmatrix} u$$

$$y = \begin{bmatrix} 1 & 2 \end{bmatrix} \begin{bmatrix} x_1 \\ x_2 \end{bmatrix}$$

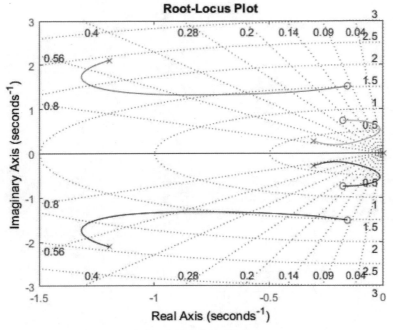

Figure 7.19 Output figure.

Find the transfer function $G(s)$ of the system using MATLAB.

Solution:

$$A = \begin{bmatrix} -4 & -1 \\ 2 & -3 \end{bmatrix}$$

$$B = \begin{bmatrix} 1 \\ 3 \end{bmatrix}$$

$$C = \begin{bmatrix} 1 & 2 \end{bmatrix}$$

The transfer function G(s) of the system is

$$G(s) = C(sI - A)^{-1}B$$

The Matlab code follows:

```
%% Matlab script
» A = [ -4 -1; 2 -3];
» B = [ 1; 3 ];
» C = [ 1 2 ];
```

```
» D = 0;
» [num, den] = ss2tf(A, B, C, D)
num =
0 7.0000 28.0000


den =
1.0000 7.0000 14.0000
```

7.4.14 Example 7.14

Determine the transfer function G(s) = Y(s)/R(s), for the following system representation in state space form.

$$\dot{x} = \begin{bmatrix} 0 & 3 & 2 & 1 \\ 3 & 1 & 0 & 2 \\ 0 & 0 & -1 & 0 \\ 3 & -2 & 0 & 1 \end{bmatrix} x + \begin{bmatrix} 0 \\ 3 \\ 2 \\ 1 \end{bmatrix} r$$

$$y = \begin{bmatrix} 1 & 1 & 0 & 3 \end{bmatrix} x$$

Response:

```
%% Matlab script
A=[ 0 3 2 1 ; 3 1 0 2 ; 0 0 -1 0 ; 3 -2 0 1 ];
B=[ 0; 3; 2; 1 ];
C = [ 1 1 0 3 ];
D = 0;

statespace = ss(A, B, C, D)

statespace =

    A =
              x1  x2  x3  x4
          x1   0   3   2   1
          x2   3   1   0   2
          x3   0   0  -1   0
          x4   3  -2   0   1


    B =
              u1
```

```
              x1     0

              x2     3
              x3     2
              x4     1

     C =

              x1    x2    x3    x4
         y1    1     1     0     3

     D =

              u1
         y1    0

Continuous-time state-space mode

[num, deg] = ss2tf(A, B, C, D)

num =

         0      6.0000    -2.0000    66.0000    -38.0000

deg =      1.0000   -1.0000   -9.0000   -7.0000    0.0000


G=tf(num, deg)

G =
              6 s^3 - 2 s^2 + 66 s - 38
         ------------------------------------
         s^4 - s^3 - 9 s^2 - 7 s + 7.853e-15

Continuous-time transfer function.
```

7.4.15 Example 7.15

Plot the step response using MATLAB for the following system represented in state space, where u(t) is the unit step.

$$\dot{x} = \begin{bmatrix} 5 & 2 & 0 \\ 0 & -10 & 2 \\ 0 & 0 & -1 \end{bmatrix} + \begin{bmatrix} 0 \\ -1 \\ 1 \end{bmatrix} u(t)$$

$$y = \begin{bmatrix} 0 & 0 & 1 \end{bmatrix} x$$

$$x(0) = \begin{bmatrix} 0 \\ 0 \\ 0 \end{bmatrix}$$

```
%% Matlab script
» A = [ -5 2 0; 0 -10 2; 0 0 -1]

A =
        -5        2        0
         0      -10        2
         0        0       -1

» B = [ 0; -1; 1]

B =
         0
        -1
         1

» C = [ 0 0 1]

C =
         0        0        1

» D = 0

D =

         0

» S = ss(A, B, C, D)

S =

A =
```

```
              x1     x2     x3
       x1     -5      2      0
       x2      0    -10      2
       x3      0      0     -1

B =
              u1
       x1      0
       x2     -1
       x3      1

C =
              x1    x2    x3
       y1      0     0     1

D =
              u1
       y1      0

Continuous-time state-space model.

» step(S)
```

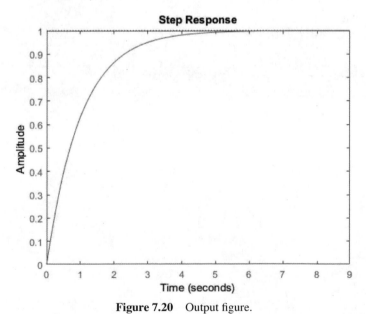

Figure 7.20   Output figure.

7.4.16 Example 7.16

A control system is defined by the following formula. Plot the four sets of Bode diagrams for the system [two for input 1, and two for input 2] using MATLAB.

$$\begin{bmatrix} \dot{x}_1 \\ \dot{x}_2 \end{bmatrix} = \begin{bmatrix} 2 & 4 \\ 2 & -6 \end{bmatrix} \begin{bmatrix} x_1 \\ x_2 \end{bmatrix} + \begin{bmatrix} 1 & 0 \\ 0 & 1 \end{bmatrix} \begin{bmatrix} u_1 \\ u_2 \end{bmatrix}$$

$$\begin{bmatrix} y_1 \\ y_2 \end{bmatrix} = \begin{bmatrix} 1 & 0 \\ 0 & 1 \end{bmatrix} \begin{bmatrix} x_1 \\ x_2 \end{bmatrix}$$

Solution:

```
There are 4 sets of Bode diagrams (2 for input 1 and 2
for input 2)


%% Matlab script
» A = [2    4; 2    -6];
» B = [1    1; 1    0];
» C = [1    0; 0    1];
» D = [0    0; 0    0];
» bode(A, B, C, D)

The plot of the four sets of Bode diagrams for the
system follows:
```

7.4.17 Example 7.17

Draw a Nyquist plot for a system defined by the following formula using Matlab.

$$\begin{bmatrix} \dot{x}_1 \\ \dot{x}_2 \end{bmatrix} = \begin{bmatrix} 2 & 4 \\ -15 & 6 \end{bmatrix} \begin{bmatrix} x_1 \\ x_2 \end{bmatrix} + \begin{bmatrix} 0 \\ 8 \end{bmatrix} \begin{bmatrix} u_1 \\ u_2 \end{bmatrix}$$

$$y = \begin{bmatrix} 1 & 0 \end{bmatrix} \begin{bmatrix} x_1 \\ x_2 \end{bmatrix} + [0]\, u$$

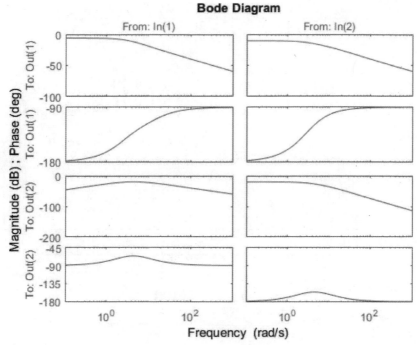

Figure 7.21   Output figure.

Solution:

Since the system has a single input *u* and *a*
single output *y*, a Nyquist plot can be obtained
by using the command **nyquist (A, B, C, D)** or
nyquist (A, B, C, D, 1).

```
%% Matlab script
» A = [ 2 4; -15 6 ];
» B = [ 0; 8 ];
» C = [ 1 0 ];
» D = [ 0 ];
» nyquist(A, B, C, D);
» grid;
» title('Nyquist plot');
```

The Nyquist plot follows:

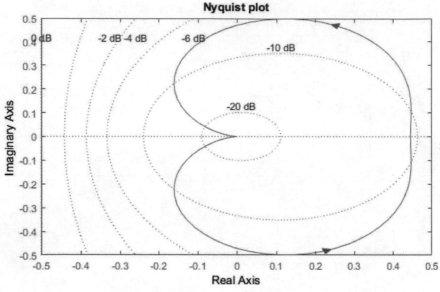

Figure 7.22   Output figure.

7.4.18 Example 7.18

Obtain the unit-step response and unit-impulse response of the following system using MATLAB

$$\begin{bmatrix} \dot{x}_1 \\ \dot{x}_2 \end{bmatrix} = \begin{bmatrix} -1 & -1.5 \\ 2 & 0 \end{bmatrix} \begin{bmatrix} x_1 \\ x_2 \end{bmatrix} + \begin{bmatrix} 1.5 \\ 0 \end{bmatrix} \begin{bmatrix} u_1 \\ u_2 \end{bmatrix}$$

$$y = \begin{bmatrix} 1 & 0 \end{bmatrix} \begin{bmatrix} x_1 \\ x_2 \end{bmatrix}$$

Solution:

```
%% Matlab script
>> %Unit-step response
>> A = [-1    -1.5 ;  2   0 ];
>> B = [ 1.5 ; 0 ];
>> C = [ 1    0 ];
>> D = [ 0 ];
```

```
>> [y, x, t] = step(A, B, C, D);
>> plot(t, y);
>> grid;
>> title('Unit-step response');
>> xlabel('t Sec');
>> ylabel('Output');

The unit-step response follows:
```

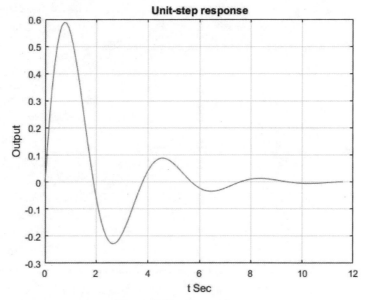

Figure 7.23 Output figure.

```
%% Matlab script
>> %Unit-impulse response
>> A = [-1    -1.5 ; 2   0 ];
>> B = [ 1.5 ; 0 ];
>> C = [ 1    0 ];
>> D = [ 0 ];
>> impulse (A, B, C, D);

The unit-impulse response follows:
```

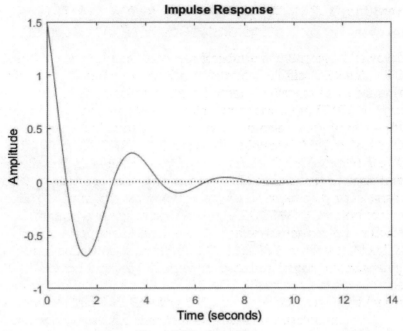

Figure 7.24   Output figure.

8

Overview of R&D Projects and Support Actions in Relevant Topics

Alkis Konstantellos[1,2]

[1]Senior C&I Engineer, Athens, Greece
[2]European Commission (retired), Complex Systems and Advanced Computing Unit, Brussels, Belgium

Abstract

This is selective set of public information of EU R&D Projects and Support Actions in Cyber-Physical Systems, Internet of Things, System of Systems, Decision Making foundations, related applications and Initiatives (funded under H2020, European Research Council and Marie Skłodowska-Curie actions). Details of the projects can be found on their websites, the road-mapping and strategy projects indicated below and in the EU portals:

- *Cordis*: http://cordis.europa.eu/projects/home_en.html
- *H2020*: https://ec.europa.eu/programmes/horizon2020/en/h2020-sections-projects
- and *ERC*: https://erc.europa.eu/projects-and-results/erc-funded-projects

This list is primarily focused on the topics discussed in this chapter and therefore it is not an exhaustive literature survey. Similarly, projects may address more than one domain, but are not repeated. Most project acronyms are in capital letters. Certain projects already completed are mentioned here because their websites still include useful reports and information on their follow-up and ongoing activities.

Disclaimer: The opinions expressed in this chapter, the selected project examples and their grouping, are those of the author and do not necessarily reflect the views of the European Commission on the topics discussed.

8.1 Road-Mapping Projects on CPS, IoT, SoS, Combined CPS–SoS and Related Technologies

CyPhERS: Cyber-Physical European Roadmap and Strategy, http://cyphers. eu/node/5 and http://www.cyphers.eu/sites/default/files/D2.2.pdf, CPSOS: Analysis of the State-of-the-Art and Future Challenges in Cyber-physical systems of systems (*completed*) – http://www.cpsos.eu/, HYCON: Hycon2: highly-complex and networked control systems (*completed*) – http:// www.hycon2.eu, SCORPIUS (European Roadmap: Cyber-Physical Systems in Manufacturing), NECS (European Network for Cyber-security), INSIGHT (In-depth support for innovation and exploitation in Smart Systems Integration), EPoSS: European technology platform on smart systems integration – http://www.smart-systems-integration.org/, LOGSEC (Development of a strategic roadmap towards a large scale demonstration project in European logistics and supply chain security), ROAD2SOS (*completed*), ROAD2CPS (Strategic action for future CPS through roadmaps, impact multiplication and constituency building); **SMEs:** EURO CPS (European Network of competencies and platforms for Enabling SME from any sector building Innovative CPS products to sustain demand for European manufacturing).

8.2 Systems Related Foundations and Novel Concepts

OPT4SMART (Distributed Optimization Methods for Smart Cyber-Physical Networks), SPHINX (Co-Evolution Framework for Model Refactoring and Proof Adaptation in Cyber-Physical Systems), ViCyPhySys (virtual Cyber-Physical Systems), SYMBIOSYS (Symbolic Analysis of Temporal and Functional Behavior of Networked Systems), FEDERATES (A Foundation for Engineering Decentralized Self-Adaptive Software Systems), CONDISHYST (Constructive and Disruptive Effects of Noise in Non-linear Systems with Hysteresis), DMASD4CA (Distributed Multi-way Analysis of Stream Data for Detection of Complex Attacks), 4DCO-GC (4 Dimension Contracts – Guidance and Control), NOUS (Probabilistic Inverse Models for Assessing the Predictive Accuracy of Inelastic Seismic Numerical Analyses), CoEvolFramework (Unified Framework for the Analysis of Co-evolutionary Systems), INTAS (Development of a unified approach and software for numerically solving inverse and optimisation problems for distributed systems), DSTOA (Descriptive set theory and operator algebras), SADCO (Sensitivity Analysis for Deterministic Controller Design), LUCRETIUS (Foundations for Software Evolution).

8.3 Cross-Layer Programming

ARGO (WCET-Aware Parallelization of Model-Based Applications for Heterogeneous Parallel Systems), CERBERO (Cross-layer model-based framework for multi-objective design of Reconfigurable systems in uncertain hybrid environments), PHANTOM (Cross-Layer and Multi-Objective Programming Approach for Next Generation Heterogeneous Parallel Computing Systems), CPS (Cross-Layer Design of Securing Positioning).

8.4 Systems of Systems and CPS

COMPASS, DANSE, AMADEOS, DYMASOS (*completed* projects), ERA-PLANET (The European network for observing our changing planet) Note Refers to GEOSS, CyberWiz (Cyber-Security Visualization and CAD-Tool for the Vulnerability Assessment of Critical Infrastructures), HyVar (Scalable Hybrid Variability for Distributed Evolving Software Systems)[1].

8.5 Control and Optimization in CPS and SoS

LOCAL4GLOBAL (*completed*), CONTROL-CPS (Reactive control protocols for cyber-physical systems), COCKPITCI (Cybersecurity on SCADA: risk prediction, analysis and reaction tools for Critical Infrastructures), DEMAND (Decentralised monitoring and control for networked dynamic systems), SAGE-CPSOC (Self-Aware CPSoCs with Hierarchical Goal Management), APROCS Automated Linear Parameter-Varying Modeling and Control Synthesis for Nonlinear Complex Systems), OCPS (Platform-aware Model-driven Optimization of CPS), Co4Robots (Achieving Complex Collabor. Missions via Decentralized control).

8.6 CPS Modeling, Design, Methods, and Tools

AXIOM (Agile, extensible, fast I/O Module for the cyber-physical era), COSSIM (A Novel, Comprehensible, Ultra-Fast, Security-Aware CPS Simulator), INTO-CPS (Integrated Tool chain for model-based design of CPSs), MAYA (Multi-disciplinary integrated simulation and forecasting tools, empowered by digital continuity and continuous real-world synchronization, towards reduced time to production and optimization), CPSWARM

---

[1]Refers to Domain Specific Variability Languages (DSVL).

(System integration and tools to support engineering of CPS swarms), SINet (Software-defined Intermittent Networking); **Mixed criticality Systems:** CERTAINTY and VIRTICAL (*completed*); DREAMS, PROXIMA, CONTREX, **Big Data:** SYSDYNET (Data-driven Modelling in Dynamic Networks), EXCELL (Smart Cyber-Physical Systems applications through exploitation of Big Data in the context of Production Control and Logistics).

8.7 CPS Security, Safety, Trust, and Testing

MODESEC (Model-based Design of Secure Cyber-Physical Systems), SAFURE (Safety and security by design for interconnected mixed-critical cyber-physical systems), SCISSOR (Security In trusted SCADA and smart-grids), SMSCOM (Self-managing situated computing) Note: ERC project, U-TEST (Testing Cyber-Physical Systems under Uncertainty: Systematic, Extensible, and Configurable Model-based and Search-based Testing Methodologies), AMASS (Architecture-driven, Multi-concern and Seamless Assurance and Certification of Cyber-Physical Systems), ENABLE-S3 (European Initiative to Enable Validation for Highly Automated Safe and Secure Systems), EATS (ETCS Advanced Testing and Smart Train Positioning System), EURO-MILS (Novel security architecture for embedded systems), CAR (Context-Active Resilience in Cyber Physical Systems), REVEN-X1: Automatic Vulnerability Detection in Binary, SHARCS (Secure Hardware-Software Architectures for Robust Computing Systems), SPARKS (Smart Grid Protection Against Cyber Attacks), SAFE CLOUD (Secure and Resilient Cloud Architecture), SAFECOP (Safe Cooperating Cyber-Physical Systems using Wireless Communication), TAPPS (Trusted Apps for open CPS), WITDOM (Empowering privacy and security in non-trusted environments); **Dependability:** CoMMiCS (Sharing resources for enhanced dependability) Note: addressing TTA, DEIS (Dependability Engineering Innovation for CPS), 5G-ENSURE (5G Enablers for Network and System Security and Resilience), TUTORIAL (Research and Innovation Capacity in Nanoelectronics Based Dependable Cyber-Physical Systems); **CPS Standardization:** CP-SETIS (Towards Cyber-Physical Systems Engineering Tools Interoperability Standardisation); **Certification:** AMASS (Architecture-driven, Multi-concern and Seamless Assurance and Certification of Cyber-Physical Systems), CERTMILS (Compositional security certification for medium-to high-assurance COTS-based systems in environments with emerging threats).

8.8 CPS Verification

ADVANCE (Advanced Design and Verification Environment for Cyber-Physical Systems Engineering), ERCPS (Unifying Control and Verification of Cyber-Physical Systems), IMMORTAL (Integrated Modelling, Fault Management, Verification and Reliable Design Environment for Cyber-Physical Systems), VESSEDIA (Verification Engineering of Safety and Security Critical Dynamic Industrial Applications).

8.9 CPS Platforms

EoT, HERCULES, LPGPU2, SAFEPOWER, TULIPP, TETRACOM, EUROCPS (European Network of competencies and platforms for Enabling SME from any sector building Innovative CPS products to sustain demand for European manufacturing), CPSELABS (CPS Engineering Labs – expediting and accelerating the realization of cyber-physical systems), PLATFORMS4CPS (Creating the CPS Vision, Strategy, Technology Building Blocks and Supporting Ecosystem for Future CPS Platforms), BEinCPPS (Business Experiments in Cyber-Physical Production Systems). Note: including CPS, BONSEYES (Platform for Open Development of Systems of AI).

8.10 CPS and Manufacturing

MANTIS (Cyber-Physical System based Proactive Collaborative Maintenance), CONNECTED FACTORIES (Industrial scenarios for connected factories), FOR ZDM (Integrated Zero Defect Manufacturing Solution for High Value Adding Multi-Stage Manufacturing systems), GOOD MAN (agent Oriented Zero Defect Multi-stage manufacturing), DCOR (Decentralised and Collaborative Production Management via Enterprise Modelling and Method Reuse).

8.11 Industry 4.0 and CPS

SMS 4.0 (Industry 4.0 for SMEs – Smart Manufacturing and Logistics for SMEs in an X-to-order and Mass Customization Environment), DISRUPT (Decentralised architectures for optimised operations via virtualised processes and manufacturing ecosystem collaboration), AceForm4.0 (Activating Value Chains for EU leadership in Formulation Manufacturing 4.0).

8.12 IoT and Underpinning Challenges

(BIG IoT) Bridging the Interoperability Gap of the Internet of Things, (UNIFY-IoT) Supporting Internet of Things Activities on Innovation Ecosystems, EoT (Eyes of Things), PREVIE (Predictive system to recommend Injection mold-setup in Wireless sensor networks), PRIME (Ultra-Low Power technologies and memory architectures for IoT), Ebbits (Enabling business-based Internet of Things and Services – An Interoperability platform for a real-world populated Internet of Things domain, C-Levitonics (Classical Levitonics: transposing quantum levitons to classical waves for single side band wireless data transmissions), U4IoT (User Engagement for Large Scale Pilots in the Internet of Things), AUTOPILOT (Automated driving Progressed by Internet Of Things), HORSE (Smart integrated Robotics system for SMEs controlled by Internet of Things based on dynamic manufacturing processes); **5G for IoT:** MAPS (Millimetre Wave Massive Arrays enabling RFID/Radar Applications on 5G Smartphones).

8.13 International Cooperation – Examples

COLUMBUS (1st EU US cooperation project, *completed*) PICASSO (ICT Policy, Research and Innovation for a Smart Society: towards new avenues in EU-US ICT collaboration), Wise-IoT (Worldwide Interoperability for Semantics IoT) Note: Refers to EU-Korea Testbeds, CPS Summit (Transatlantic US-EU CPS Summit), TAMS4CPS (Trans-Atlantic Modelling and Simulation For Cyber-Physical Systems), T-AREA SoS (*completed*).

8.14 Decision Making (Methods Applied)

PRISMS (The Privacy and Security Mirrors: "Towards a European framework for integrated decision making"), STARR (Decision Support and self-management system for stroke survivors), IDMSEBE (Intelligent Decision Making Systems in European Business and Economics), POWER-OM (Power consumption driven Reliability, Operation and Maintenance optimisation), FLOURISH (Aerial Data Collection and Analysis, and Automated Ground Intervention for Precision Farming), ASSET (Analysing and Striking the Sensitivities of Embryonal Tumours)[2], GEO SAFE (Geospatial based Environment for Optimisation Systems Addressing Fire Emergencies).

---

[2]Developing reusable code for conducting Markov chain Monte Carlo (MCMC) inference.

8.15 Decision Processes – Human in the Loop

ROBUST SENSE (Robust and Reliable Environment Sensing and Situation Prediction for Advanced Driver Assistance Systems and Automated Driving), RESOLUTE (Resilience management guidelines and Operationalization applied to Urban Transport Environment); DBCAR (Software for driverless vehicles), LIVCODE (Life-like visual information processing for robust collision detection), CASCADE (Model-based Cooperative and Adaptive Ship-based Context Aware Design), TELEPRESENCE SURGERY (Human-in-the-Loop Telepresence Control for robot-assisted surgery), FMHAI (Formal Analysis and Modelling of Human-Automation Interaction), VINBOT (Autonomous Cloud-based Vineyard Robot to optimise yield and wine quality).

8.16 Collision Avoidance (CA) Including ACAS-X

ARTRAC (Advanced Radar Tracking and Classification for Enhanced Road), SARNET2 (Severe Accident Research Network of Excellence 2), MYCOPTER (Enabling Technologies for Personal Air Transport Systems), ULTRA (Unmanned Aerial Systems in European Airspace) Note addressing ACAS-X, REWARD (Real Time Wide Area Radiation Surveillance System); PJ11 CAPITO (Enhanced Air and Ground Safety Nets)[3].

8.17 Concluding Comments

This chapter presented titles of some 142 R&D projects, funded by the European Commission (on-going/under development, or already completed). The main selection criterion was the relevance to the topics (and the work of the related research teams) of this book and in particular to the introductory chapters. The rough categorisation may not be strict, but hopefully useful to facilitate to locate interesting activities. The focus was primarily on European projects [1, 9]. In Europe, research is also funded in many more related topics, for example under robotics [3]. Furthermore, as an indication of the enormous international interest in governmental funded R&D in the areas addressed in this book, the reader may scan the extensive regular project activities, for example by NSF in the USA [4], DFG under GEPRIS in Germany [5], OECD [6], Russia [7], India [8] and additional project information sources

---

[3]Eurocontrol activity in CA.

are available such as in ECSEL [10] and ARTEMIS Industrial Association [2, 11]. Educational and Training issues in these fields are of paramount importance for the establishment of the next generation S&T communities, but go beyond the scope of this chapter. Finally, since R&D activities continue around the world, the current lists will be superseded in the next years, but hoping that these projects will have added value to the existing portfolio of methods, tools and applications in the systems domains.

Acknowledgement

The author would like to thank his ex-colleague Dr. Werner Steinhögl, European Commission, Brussels, Cyber-Physical Systems area, (recently Head of Sector in Industrial Laser-based Technologies, Photonics Unit), for his kind comments on an early draft of this chapter and his useful discussions on R&D issues in systems, computing and applications.

References

[1] http://ec.europa.eu/research/index.cfm
[2] https://artemis-ia.eu/ecsel-call.html
[3] http://cordis.europa.eu/search/result_en?q=robotics&searchType=simple
[4] https://www.nsf.gov/awardsearch/
[5] http://gepris.dfg.de/gepris/OCTOPUS;jsessionid=0E47FCB5C97F0287
    C1D66987A9ACDC93
[6] https://www.oecd.org/sti/outlook/e-outlook/stipolicyprofiles/competenc
    estoinnovate/publicresearchpolicy.htm
[7] http://government.ru/en/docs/2129/
[8] http://www.serb.gov.in/irrd.php
[9] http://www.mscrnd.com/horizon-2020-2018-2020-work-programme-
    indications/
[10] http://www.ecsel-ju.eu/web/index.php
[11] https://artemis-ia.eu/projects-1.html

Index

About the Editors

Kostas Siozios received his Diploma, Master and Ph.D. Degree in ECE from the Democritus University of Thrace, Greece, in 2001, 2003 and 2009, respectively. He is currently an Assistant Professor in School of Physics of Aristotle University of Thessaloniki (A.U.Th.), Thessaloniki, Greece. From 2009–2016, he was senior researcher in the National Technical University of Athens (N.T.U.A.), Greece. His research interests are in the areas of reconfigurable and embedded systems, the development of Computer-Aided Design (CAD) algorithms, the cyber-physical systems, as well as the implementation of low-complexity control mechanisms. He has published more than 150 papers in international journals and conferences, while he is also co-author or co-editor of 7 books of Springer and CRC. Starting from 2002 he has participated as principal investigator in more than 20 research projects funded from the European Commission (E.C.), European Space Agency (E.S.A.), as well as the Greek Government and Industry.

Dimitrios Soudris is an Assoc. Professor in School of ECE of National Technical University of Athens, Greece. His research interests include embedded systems design, reconfigurable architectures, network-on-chip architectures and low-power VLSI design. He has published more than 400 papers in international journals and conferences. He is author and editor in eight books of Kluwer and Springer. His research work has been cited >2400 times. He is the Head of Embedded Systems group consisting of four Senior Investigators (Post-Docs) and eight Ph.D. students and a number of M.Sc. students. He is leader and principal investigator in numerous research projects (>45) funded from the European Commission, ENIAC-JU, European Space Agency and the Greek Government and Industry. He served as General Chair and Program Chair for PATMOS and General Chair of IFIP-VLSI-SOC 2008, PARMA 2011 & 2013 Workshop, SAMOS 2015, HiPEAC CSW 2014. He received an award from INTEL and IBM for the project results of LPGD #25256 and awards from Int. Conf. VLSI 2005 and ASP-DAC 05 for the results of the project AMDREL IST-2001-34379 and recently the HiPEAC Award for a DAC '10 paper for the project MNEMEE FP7-216224 and DAC '13 for the project ENIAC-TOISE-282557-2.

Elias Kosmatopoulos received the Diploma, M.Sc. and Ph.D. degrees from the Technical University of Crete, Greece, in 1990, 1992, and 1995, respectively. He is currently a Professor with the Department of Electrical and Computer Engineering, Democritus University of Thrace, Xanthi, Greece. Previously, he was a faculty member of the Department of Production Engineering and Management, Technical University of Crete (TUC), Greece, a Research Assistant Professor with the Department of Electrical Engineering-Systems, University of Southern California (USC) and a Postdoctoral Fellow with the Department of Electrical & Computer Engineering, University of Victoria, B.C., Canada. Dr. Kosmatopoulos' research interests are in the areas of adaptive optimization and control, IoT and Cyper-Physical Systems and their applications to energy efficient buildings, smart grids, robotics and intelligent transportation systems. He is the author of over 55 journal papers. He serves as an Associate Editor for IEEE Trans. on Intelligent Transportation Systems. He has been leading many research projects funded by the European Union with a total budget of about 20 Million Euros.